Natural Computing Series

Series Editors: G. Rozenberg
Th. Bäck A.E. Eiben J.N. Kok H.P. Spaink

Leiden Center for Natural Computing

For further volumes:
http://www.springer.com/series/4190

Thomas Bäck • Christophe Foussette
Peter Krause

Contemporary Evolution Strategies

 Springer

Thomas Bäck
Leiden University
Leiden
The Netherlands

Christophe Foussette
Peter Krause
divis intelligent solutions GmbH
Dortmund
Germany

Series Editors
G. Rozenberg (Managing Editor)

Th. Bäck, J.N. Kok, H.P. Spaink
Leiden Center for Natural Computing
Leiden University
Leiden, The Netherlands

A.E. Eiben
Vrije Universiteit Amsterdam
The Netherlands

ISSN 1619-7127 Natural Computing Series
ISBN 978-3-642-40136-7   ISBN 978-3-642-40137-4 (eBook)
DOI 10.1007/978-3-642-40137-4
Springer Heidelberg New York Dordrecht London

Library of Congress Control Number: 2013950140

Printed on acid-free paper

Springer is part of Springer Science+Business Media (www.springer.com)

Contents

List of Figures

List of Algorithms

Chapter 1
Introduction

1.1 Background and Motivation

Today, in 2013, when this small book is being written, *evolutionary algorithms* are established as a well-known and widely used class of heuristics, inspired by the model of organic evolution, for solving optimization problems. And this really means that these algorithms are regularly used in real-world applications, and some algorithmic variants have been incorporated into standardized off-the-shelf software toolboxes. Between 1990—when the first author of this book entered into the field which was known under the term "genetic algorithms" only—and today, the field has seen tremendous development and has earned enormous scientific recognition. The generalization and unification of algorithms, interpreting and formalizing genetic algorithms, evolution strategies, and evolutionary programming as instantiations of a more generic concept of so-called evolutionary algorithms [8] has been an important step towards understanding more general principles and exchanging algorithmic concepts between specific algorithms. The concept of strategy parameter self-adaptation is a specific component of evolution strategies, which was originally introduced and is still in use within the original (μ, λ)-evolution strategy introduced by Schwefel [62], also described in [8]. Since about 1994, numerous extensions of this concept have addressed the topics of derandomization and covariance matrix adaptation in particular, which has resulted in the development of the contemporary evolution strategies, such as the covariance matrix adaptation strategy or CMA-ES [31]. The aim of this book is to summarize these developments and present them in a concise way, so that the reader can get an overview of the diverse range of developments in this field. In addition to pseudo-code representations of the corresponding algorithms, source code for non-commercial use is also provided for the majority of the algorithms.

As far as this book is concerned, the core application field of interest for evolution strategies is optimization tasks with a very limited budget of objective function evaluations. Loosely speaking, we can think of an optimization problem as a black box which, for a given input parameter vector, generates an associated

T. Bäck et al., *Contemporary Evolution Strategies*, Natural Computing Series, DOI 10.1007/978-3-642-40137-4_1, © Springer-Verlag Berlin Heidelberg 2013

scalar output value, and we are interested in finding an assignment of values of input parameters which maximizes or, equivalently, minimizes, the scalar output value. Optimization is a ubiquitous task, relevant to industry, government, and society as a whole, as we will see below in a few illustrative examples.

Evolution strategies, the focus topic of this book, form a subclass of evolutionary algorithms (see, e.g., [9, 10] for a complete coverage of the field), mainly designed for optimization problems defined over vectors of floating point numbers as input parameters, i.e., for continuous parameter optimization tasks. Historically, all variations of evolution strategies can be characterized by the fact that they use a mutation operator based on a multivariate normal distribution and rely on a special approach for online adaptation of the variances (and covariances) of the normal distribution. This latter approach is typically called *self-adaptation*.

Based on hundreds of industrial optimization tasks we have worked on during our professional lives, we have experienced the enormous capabilities of evolution strategies in the real world. Problems range from engineering design applications (e.g., how to optimize the geometric two- or three-dimensional design of an airfoil for maximum lift or minimum drag) to logistics and scheduling applications (e.g., how to choose shipper batch sizing and sequencing to achieve optimal throughput at minimum energy costs within a pipeline network) and further on to food production (e.g., how to tune production process parameters so as to maximize shelf-life stability of chocolate), just to mention a few examples. Common to these three examples, like many other instances of optimization tasks, is the fact that no closed-form analytical description of the function mapping input parameter vectors to objective function values exists—therefore, traditional analytical methods are not applicable, and also gradient information and even the Hessian are not available. Such tasks are also called *black-box-optimization* tasks, since the mapping of input vectors into objective function values is essentially unknown, i.e., a black box. Often, this black box is represented by a simulation model, executing a simulation of a real-world system on a computer (e.g., simulating the noise, vibration and harshness behavior of a car body), and sometimes it is even a real-world experiment, when no simulation is available at all (e.g., producing a batch of chocolate and then measuring its stability duration). Besides the black box aspect, many real-world optimization problems are characterized by the fact that the objective function evaluation by means of simulation or even by an experiment can be quite time consuming and therefore expensive, so that real-world optimization is often looking for the impossible: delivering a solution as good as possible *and* as fast as possible (i.e., with as few objective function evaluations as possible).

The power of evolution strategies is really their ability to combine these two conflicting goals—using few objective function evaluations and delivering high-quality solutions. They achieve this through small population sizes (which is typically seen to be uncommon in evolutionary computation) and the ability to self-adapt the probability density function used for generating mutations. Since our professional focus has always been on this key feature of evolution strategies, the empirical investigation presented in Chap. 5 of this book focuses on comparing the performance of algorithms by looking at a relatively small[1] number of function evaluations only.

Based on the model of the Darwinian theory of evolution [19], evolutionary algorithms including evolution strategies use principles from evolutionary biology for solving black-box-optimization problems. The individuals in an evolution strategy consist of vectors of input parameters to the objective function and of strategy parameters for the mutation operator, so-called endogenous parameters. Multiple individuals together form a population, which undergoes an evolution cycle during each iteration (also called generation) of the algorithm. Through randomized processes of recombination and mutation, offspring individuals are generated from the current parent population, and the objective function values (often also called fitness values) of the offspring individuals are calculated by applying the black box (i.e., by running the simulation program with the input parameter values specified in the individuals). Finally, the selection operator chooses those individuals with best fitness values to become parents of the next generation.

Since the introduction of the first evolution strategy, the so-called (1+1)-ES, by Rechenberg and Schwefel [51, 52, 59] in the 1960s and 1970s, many variants of the algorithm have been developed. In particular, variations were developed in the 1990s which use a so-called *derandomization* approach for the strategy parameter adaptation mechanism, characterizing a major shift of paradigm for evolution strategies. The derandomization efforts ultimately led to the development of the so-called CMA-ES (*covariance matrix adaptation evolution strategy*), which adapts a full covariance matrix in a derandomized way [31]. The derandomized variants, and in particular the CMA-ES and its successors, are the algorithms which are called *contemporary evolution strategies* throughout this book, as they constitute a new era in the field of evolution strategies.

A key goal of this book is to provide the reader with an overview of these modern evolution strategy variants, as there is no book yet where all of them are described in a unified context. Moreover, we also provide a taxonomy of these algorithms with respect to a number of different characteristics, and we use this taxonomy to provide recommendations for the practical usage of the CMA-ES and its successors.

Most publications about modern evolution strategies compare those algorithms either with each other, or with some competitor from a different class of evolutionary or other optimization algorithms, typically by using a set of standardized test functions. These empirical comparisons are typically based on a large number of objective function evaluations.[2] As explained above, however, many typical real-world application cases do not allow for such an enormous number of function evaluations, simply because of the prohibitively large effort in terms of time

---

[1] As the reader will realize, the numbers used here are still much larger than what would often be available in real-world applications (up to a few hundreds), but they are much smaller than what is typically used for comparing algorithms.

[2] For *Black-Box-Optimization Benchmarking* (BBOB) [34], the recommended number of function evaluations is $10^6 n$, for n-dimensional test problems.

and/or money for even a single objective function evaluation. For these cases, it is very important to understand the behavior of modern evolution strategies (and optimization algorithms in general) when only a small number of function evaluations are permitted, and to identify the best-performing algorithm with respect to this constraint on the number of evaluations. In principle, it could even be possible to identify completely different convergence behavior for evolution strategy variants for a large number of function evaluations as opposed to a small number, yielding a different winning strategy in the two cases. This is exactly the question we focus on in the empirical results reported in this book, i.e., the convergence behavior of modern evolution strategies for a relatively small number of permitted function evaluations.

1.2   Structure of the Book

In the following, we give a brief outline of the structure of this book.

In Chap. 2, the main chapter of the book, variants of modern evolution strategies are described by using pseudocode notation and explaining the algorithms in detail. The chapter starts by defining the optimization problem, introducing evolution strategies as a specialization of evolutionary algorithms (for the interested reader, [8] provides an integrated view of evolutionary algorithms as a generic paradigm for optimization), and describing the concept of mutation in the n-dimensional continuous search space $\mathbb{R}^n$ (see Sect. 2.1). Then, the evolution strategy variants are introduced, following the chronological order of their publication, split into modern evolution strategies and a small selection of older evolution strategies (i.e., before the inception of the CMA-ES).

Chapter 3 then provides a taxonomy of modern evolution strategies, characterizing them by thematic areas as identified in Sect. 3.1 and by interesting features introduced in Sect. 3.2. Based on their description and their taxonomy classification, recommendations for the practical usage of modern evolution strategies are provided in Sect. 3.3.

Chapter 4 then presents a systematic empirical investigation of modern evolution strategies based on performance measures as introduced in Sect. 4.1, a careful experimental setup and execution of experiments as described in Sect. 4.2, and a final presentation and discussion of the results, given in Sect. 4.3.

Finally, Chap. 5 summarizes some of the key findings and takes a short look at the future of evolution strategy research as we perceive it—through the eyes of the practitioner interested in solving extremely complex optimization tasks.

1.3   Notation

The key notational elements as used throughout this book are summarized in Table 1.1.

Table 1.1 Notation

Symbol	Description
$\mathbf{x}$	Vector $\mathbf{x}$
$\mathbf{A}$	Matrix $\mathbf{A}$
$\mathbf{I}$	Unity matrix, i.e., neutral element w.r.t. multiplication
$\text{diag}(\mathbf{d})$	Diagonal matrix $\mathbf{D} = \text{diag}(\mathbf{d})$ with vector $\mathbf{d}$ of diagonal elements
$\mathbf{v}^T, \mathbf{A}^T$	Transposition of vector $\mathbf{v}$ or matrix $\mathbf{A}$
x_i	Indexed component of a vector $\mathbf{x} = (x_1, \ldots, x_n)^T \in \mathbb{R}^n$
$A_{i,j}$	Indexed component of a matrix $\mathbf{A} \in \mathbb{R}^{n \times m}$
$\Psi.\text{age}$	Component notation, as in object-oriented programming, age is a component of Ψ
$\leftarrow$	Assignment
$\mathbf{x}_{i:m}$	Indexed access to a sorted set $S = \{\mathbf{x}_1, \ldots, \mathbf{x}_m\}$: $\mathbf{x}_{i:m}$ is $\mathbf{x} \in S$ with rank i
$\text{tr}(\mathbf{A})$	Trace of matrix $\mathbf{A} \in \mathbb{R}^{n \times n}$: $\text{tr}(\mathbf{A}) = \prod_{i=1}^n A_{i,i}$
$\mathbf{u} \otimes \mathbf{v}$	Element-wise multiplication of vectors $\mathbf{u}, \mathbf{v} \in \mathbb{R}^n$: $\mathbf{u} \otimes \mathbf{v} = \mathbf{w}$ where $\mathbf{w} \in \mathbb{R}^n$ and $w_i = u_i \cdot v_i$ for $i \in \{1, \ldots, n\}$
$\langle \mathbf{x} \rangle$	We use $\langle \mathbf{x} \rangle$ to denote a single parent; the brackets $\langle \rangle$ do not denote an operator

1.4   Source Code

The Octave source code (proprietary implementations) of most of the algorithms compared in this book can be downloaded for non-commercial use only. The following algorithms are available:

- $(1 + 1)$-ES
- (μ, λ)-MSC-ES
- DR1
- DR2
- DR3
- LS-CMA-ES
- $(1 + 1)$-Cholesky-CMA-ES
- (μ, λ)-CMSA-ES
- sep-CMA-ES
- $(1 + 1)$-Active-CMA-ES

In order to download the code, contact information must be provided. Download link: http://www.divis-gmbh.com/es-software.html

Chapter 2
Evolution Strategies

Prior to introducing the particular algorithms in Sect. 2.2, the more general founda-
tions of evolution strategies are introduced in Sect. 2.1. To start with, the definition
of an optimization task as used throughout this book is given in Sect. 2.1.1.
Following [58], Sect. 2.1.2 presents a discussion of evolution strategy metaheuristics
as a special case of evolutionary algorithms. In particular, the components of such
a metaheuristic—namely recombination, mutation, evaluation and selection—are
described in a general way. Due to the particular importance[1] of the mutation
operator for evolution strategies (in $\mathbb{R}^n$), it is discussed in quite some detail in
Sect. 2.1.3.

2.1 Introduction

2.1.1 Optimization

Evolution strategies are particularly well suited (and developed) for nonlinear
optimization tasks, which are defined as follows (see e.g. [17], Sect. 18.2.1.1):

$$f(\mathbf{x}) = \text{min! for } \mathbf{x} \in \mathbb{R}^n \text{ where} \tag{2.1}$$

$$g_i(\mathbf{x}) \leq 0, i \in I = \{1, \dots, m\}, h_j(\mathbf{x}) = 0, j \in J = \{1, \dots, r\}, \tag{2.2}$$

---

[1]This statement, however, is not meant to support the myth mentioned explicitly by Rudolph [58]:
"Since early theoretical publications mainly analyzed simple ES without recombination, somehow
the myth arose that ES put more emphasis on mutation than on recombination: This is a fatal
misconception! Recombination has been an important ingredient of ES from the early beginning
and this is still valid today."

T. Bäck et al., *Contemporary Evolution Strategies*, Natural Computing Series,
DOI 10.1007/978-3-642-40137-4_2, © Springer-Verlag Berlin Heidelberg 2013

and the set

$$M = \{\mathbf{x} \in \mathbb{R}^n : g_i(\mathbf{x}) \leq 0, \forall i \in I, h_j(\mathbf{x}) = 0, \forall j \in J\} \qquad (2.3)$$

is called the set of feasible points and it defines the search space of the optimization problem. A point $\mathbf{x}^* \in \mathbb{R}^n$ is called a global minimum, if

$$f^* = f(\mathbf{x}^*) \leq f(\mathbf{x}) \text{ for all } \mathbf{x} \in M \qquad (2.4)$$

Conversely, it is called a local minimum if the above inequality only holds for $\mathbf{x}$ within an ϵ-environment $U_{\epsilon(x)} \subseteq M$.

Formulating an optimization problem as a minimization task is equivalent to searching for a maximum or for a given target value, since maximization of f can be replaced by minimization of $-f$ and a target value $\bar{f}$ can be attained by minimizing $\rho(\bar{f}, f)$ with an arbitrary distance measure[2] ρ.

In this definition of an optimization task it is completely sufficient if the codomain is completely ordered, so that the inequality in Eq. 2.4 can be applied. Throughout this book, we will always deal with the codomain $\mathbb{R}$ only. Moreover, we will not explicitly deal with the handling of constraints (e.g., as defined by Eq. 2.2), and refer the interested reader to Sect. 2.3 where literature references point to state-of-the-art techniques in constraint handling. A special case of constraints are so-called box constraints, as defined below:

$$g_1(\mathbf{x}) = \mathbf{l} - \mathbf{x} \leq \mathbf{0} \text{ where } \mathbf{l} = (l_1, \ldots, l_n)^T \in \mathbb{R}^n$$
$$g_2(\mathbf{x}) = \mathbf{x} - \mathbf{u} \leq \mathbf{0} \text{ where } \mathbf{u} = (u_1, \ldots, u_n)^T \in \mathbb{R}^n \qquad (2.5)$$

Vectors $\mathbf{l}$ and $\mathbf{u}$ are called lower and upper bounds, respectively. Box constraints restrict the search space to the hyperrectangle $[l_1, u_1] \times \ldots \times [l_n, u_n]$ and are taken into account for the implementation of algorithms described in this book.

In the field of evolutionary algorithms, the vector $\mathbf{x}$ is often called the decision vector (and its parameters decision parameters), and its objective function value $f(\mathbf{x})$ is also called the fitness value.

2.1.2 Evolution Strategies as a Specialization of Evolutionary Algorithms

Following [8] and [58], evolution strategies are described here as a specialization of evolutionary algorithms. The general framework of an evolutionary algorithm is presented in Algorithm 2.1. During initialization, the first generation, consisting of

---

[2]See Sect. 12.2.1 in [17] for the definition of a distance measure.

Algorithm 2.1 General outline of an evolutionary algorithm

Initialization
repeat
    Recombination
    Mutation
    Evaluation
    Selection
until Termination criterion fulfilled

---

one or more individuals, is created, and the fitness of its individuals is evaluated. After initialization, the so-called evolution loop is entered, which consists of the operators recombination, mutation, evaluation and selection. Recombination creates new individuals, also called offspring, from the parent population. Two major types of recombination, dominant and intermediate recombination, are typically distinguished: In dominant recombination, a property of a parent individual is inherited by the offspring, i.e., this property dominates the corresponding property of the other individuals. For intermediate recombination, the properties of all individuals are taken into account, such that, e.g., in the simplest case, their mean value is used.

The mutation operator provides the main source of variation of offspring in an evolution strategy. Based on sampling random variables, properties of individuals are modified. The newly created individuals are then evaluated, i.e., their fitness values are calculated. Based on these fitness values, selection identifies a subset of individuals which form the new population which is used in the next iteration of the evolution loop. The loop is terminated based on a termination criterion set by the user, such as reaching a maximum number of evaluations, reaching a target fitness value, or stagnation of the search process.

According to [58], evolution strategies as a specific instantiation of evolutionary algorithms are characterized by the following four properties:

- Selection of individuals for recombination is unbiased.
- Selection is a deterministic process.
- Mutation operators are parameterized and therefore they can change their properties during optimization.
- Individuals consist of decision parameters as well as strategy parameters.[3]

The generic framework of an evolutionary algorithm then specializes into a $(\mu/\rho, \kappa, \lambda)$-ES,[4] as described in detail in Algorithm 2.2. Recombination and mutation are summarized here under the term variation. In addition to the description

---

[3]In the case of the $(1+1)$-ES the strategy parameters may be assigned to the algorithm itself instead of the individual, because only one set of strategy parameters is needed. This also holds for any strategy parameters which are not needed on the individual level (for example the covariance matrix of the CMA-ES).

[4]Algorithm 3 in [58].

Algorithm 2.2 $(\mu/\rho, \kappa, \lambda)$-ES

Initialization of $P^{(0)}$ with μ individuals

$\forall p \in P^{(0)} : p.\Psi.Age \leftarrow 1, p.f \leftarrow f(p.\mathbf{x})$

$t \leftarrow 0$

repeat

    $Q^{(t)} \leftarrow \emptyset$

    for $i = 1 \rightarrow \lambda$ **do**

        Sample ρ parents $p_1, \ldots, p_\rho \in P^{(t)}$ uniformly at random

        $q \leftarrow \text{Variation}(p_1, \ldots, p_\rho, \Psi_V)$

        $q.\Psi.Age \leftarrow 0, q.f \leftarrow f(q.\mathbf{x})$

        $Q^{(t)} \leftarrow Q^{(t)} \cup \{q\}$

    end for

    $P^{(t+1)} \leftarrow$ Selection of the μ best individuals from $Q^{(t)} \cup \{p \in P^{(t)} : p.\Psi.Age < \kappa\}$

    Update Ψ_V

    $\forall p \in P^{(t+1)} : p.\Psi.Age \leftarrow p.\Psi.Age + 1$

    $t \leftarrow t + 1$

until Termination criterion fulfilled

---

given in [58] (Algorithm 3), the variation operator of a $(\mu/\rho, \kappa, \lambda)$-ES is defined here by means of a parameter set Ψ_V, and the evaluation operator is explicitly mentioned. A population at generation $t \geq 0$ is denoted $P^{(t)}$ and is a set of individuals. An individual $p \in P^{(t)}$ is a tuple $(\mathbf{x}, \Psi)$ for $\mathbf{x} \in M \subseteq \mathbb{R}^n$, with M as in Eq. 2.3. The sets Ψ and Ψ_V are arbitrary finite sets representing the strategy parameters. Since these parameters are modified internally during execution of the algorithm, they are called endogenous strategy parameters. The number of parent individuals is denoted as μ, the number of offspring individuals as λ, and ρ denotes the number of parents taken into account for generating a single offspring by means of recombination. For these parameters, $\mu, \rho, \lambda \in \mathbb{N}$ and $\rho \leq \mu$ holds.

$\kappa \in \mathbb{N} \cup \{\infty\}$ represents the largest age which can be reached by any individual in the population. In contrast to endogenous parameters, μ, ρ, λ und κ are to be set by the user of the algorithm, such that they are called exogenous strategy parameters.

The setting of κ has a direct impact on the selection operator. Usually, either $\kappa = 1$ (one generation maximum lifetime) or $\kappa = \infty$ (infinite maximum lifetime) is used. The former case is also called comma-selection, the latter plus-selection. Using the standard notation of evolution strategies, this is expressed as $(\mu/\rho, \lambda)$-ES and $(\mu/\rho+\lambda)$-ES, so that κ is not explicitly stated any more. Using $\kappa < \infty$ requires the condition $\lambda \geq \mu$ to hold.

2.1.3 Mutation in $\mathbb{R}^n$

2.1.3.1 The Multivariate Normal Distribution

In [58], three guiding principles for the design of mutation operators are introduced, namely:

- Any point of the search space needs to be attainable with probability strictly larger than zero by means of a finite number of applications of mutation.
- Mutation should be *unbiased*, which can be achieved by using a *maximum entropy distribution*.[5]
- The operator is parameterized, such that the extent of variation can be controlled.

In $\mathbb{R}^n$, these requirements are fulfilled by a multivariate normal distribution. An n-dimensional random vector $\mathbf{X}$ is multivariate normally distributed with expectation $\bar{\mathbf{x}} \in \mathbb{R}^n$ and positive definite[6] covariance matrix $\mathbf{C} \in \mathbb{R}^{n \times n}$ if its probability density function is defined according to:

$$f_{\mathbf{X}}(\mathbf{x}) = \frac{1}{(2\pi)^{\frac{n}{2}} (\det \mathbf{C})^{\frac{1}{2}}} \exp\left(-\frac{1}{2}(\mathbf{x} - \bar{\mathbf{x}})^T \mathbf{C}^{-1}(\mathbf{x} - \bar{\mathbf{x}})\right) \tag{2.6}$$

(see p. 86 in [28]). In short notation, this is typically written as $\mathbf{X} \sim N(\bar{\mathbf{x}}, \mathbf{C})$, where $N(\bar{\mathbf{x}}, \mathbf{C})$ denotes the multivariate normal distribution in its general form. In mathematical equations, $N(\bar{\mathbf{x}}, \mathbf{C})$ is sometimes used like a vector, meaning a vector which is actually sampled according to the distribution given. In other words, instead of writing $\mathbf{x}' = \mathbf{x} + \mathbf{X}$ where $\mathbf{X} \sim N(\mathbf{0}, \mathbf{C})$, it is also possible to simply write $\mathbf{x}' = \mathbf{x} + N(\mathbf{0}, \mathbf{C})$.

Due to the positive definiteness of the covariance matrix $\mathbf{C}$, the following eigendecomposition exists (Theorem 1a in [58]):

$$\mathbf{C} = \mathbf{BD}^2\mathbf{B}^T \tag{2.7}$$

Here, $\mathbf{B}$ denotes an orthogonal matrix,[7] the columns of which are the eigenvectors of $\mathbf{C}$. In [29], $N(\bar{\mathbf{x}}, \mathbf{C})$ is reduced to the distribution $N(\mathbf{0}, \mathbf{I})$ by means of the eigendecomposition given in Eq. 2.7, according to:

$$N(\bar{\mathbf{x}}, \mathbf{C}) \sim \bar{\mathbf{x}} + \mathbf{BD}N(\mathbf{0}, \mathbf{I}) \tag{2.8}$$

In the field of evolution strategies, the three special cases $N(\mathbf{0}, \mathbf{I})$, $N(\mathbf{0}, \text{diag}(\delta^2))$ and $N(\mathbf{0}, \mathbf{C})$ are used for the definition of the most common algorithms. Figure 2.1 provides a sketch of the corresponding mutation ellipsoids, i.e., isolines of the probability density functions, embedded in a hypothetical two-dimensional fitness function.

The simplest case of generating the mutation $\mathbf{x}'$ from $\mathbf{x}$ is based on using $\mathbf{B} = \mathbf{I}$ and $\mathbf{D} = \sigma\mathbf{I}$ with a global step size $\delta \in \mathbb{R}^+$ for matrices $\mathbf{B}$ and $\mathbf{D}$ as used in Eq. 2.8.

---

[5]The normal distribution achieves maximum entropy among the distributions on the real domain. (See [64] for more details.)

[6]A symmetric matrix $\mathbf{A} \in \mathbb{R}^{n \times n}$ is positive definite iff $\mathbf{x}^T \mathbf{A}\mathbf{x} > 0$ for all $\mathbf{x} \in \mathbb{R}^n \setminus \{\mathbf{0}\}$ [17].

[7]For an orthogonal matrix $\mathbf{A}$, $\mathbf{AA}^T = \mathbf{A}^T\mathbf{A} = \mathbf{I}$ holds.

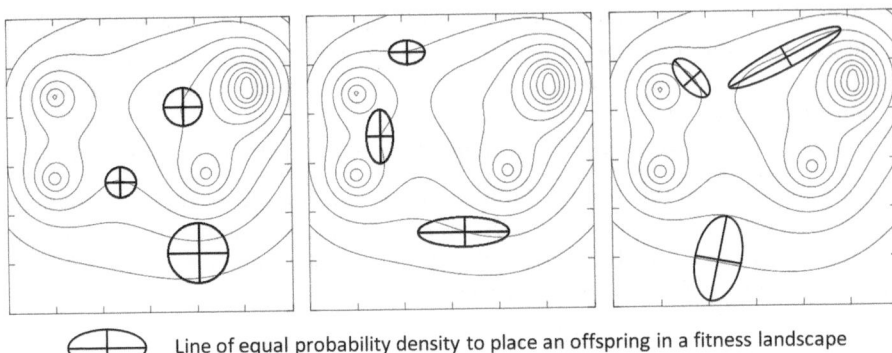

Line of equal probability density to place an offspring in a fitness landscape

Fig. 2.1 Mutation ellipsoids representing $N(\mathbf{0}, \mathbf{I})$, $N(\mathbf{0}, \text{diag}(\delta^2))$ and $N(\mathbf{0}, \mathbf{C})$ (from *left* to *right*)

$$\mathbf{x}' = \mathbf{x} + \delta \cdot N(\mathbf{0}, \mathbf{I}) \tag{2.9}$$

This corresponds with spheres with individual radii defined by δ, as indicated in the left part of Fig. 2.1. This case of an offspring distribution is called isotropic.

To turn the spheres into anisotropic ellipsoids with main axes parallel to the coordinate axes, as shown in the middle of Fig. 2.1, matrix $\mathbf{D}$ in Eq. 2.8 must be turned into a diagonal matrix $\delta = (\delta_1, \ldots, \delta_n)^T \in \mathbb{R}^n$ with different entries on the main diagonal. As in the previous case, $\mathbf{B}$ is a diagonal matrix:

$$\mathbf{x}' = \mathbf{x} + \mathbf{I}\text{diag}(\delta)N(\mathbf{0}, \mathbf{I})$$
$$= \mathbf{x} + N(\mathbf{0}, \text{diag}(\delta^2)) \tag{2.10}$$

The length ratios of the main axes of the mutation ellipsoids depend on the ratios between corresponding components of the vector δ. A rotation of mutation hyperellipsoids with respect to the coordinate axes, as shown in the rightmost part of Fig. 2.1, is achieved by using a covariance matrix $\mathbf{C}$ with off-diagonal entries different from zero. This case is denoted by the term correlated mutation. In contrast with the two previous cases, the matrix $\mathbf{B}$ is not just an identity matrix:

$$\mathbf{x}' = \mathbf{x} + \mathbf{B}\text{diag}(\delta)N(\mathbf{0}, \mathbf{I})$$
$$= \mathbf{x} + \mathbf{B}N(\mathbf{0}, \text{diag}(\delta^2))$$
$$= \mathbf{x} + N(\mathbf{0}, \mathbf{C}) \tag{2.11}$$

The choice of one of the three cases explained above has a direct impact on the complexity of the endogenous parameters controlling the multivariate normal distribution. In general, if n denotes the dimensionality of the search space, the number of endogenous strategy parameters in case of Eq. 2.9 is $O(1)$, i.e., constant. In case of 2.10 a vector of size $O(n)$ of endogenous parameters is required, and adaptation of an arbitrary covariance matrix, i.e., a symmetric $n \times n$-matrix, according to Eq. 2.11, requires $O(n^2)$ endogenous parameters.

For defining algorithm DR3 in Sect. 2.2.1 and for all algorithms based on the CMA-ES, the so-called *line distribution* [31] is of special interest: For $\mathbf{u} \in \mathbb{R}^n$, the distribution $N(\mathbf{0}, \mathbf{u}\mathbf{u}^T)$ is a multivariate normal distribution with the variance $\|\mathbf{u}\|^2$ in the direction of the vector $\mathbf{u}$. It is the normal distribution with highest probability of generating $\mathbf{u}$.

2.1.3.2 Relationship Between Covariance Matrix and Hessian

In the previous section, using a multivariate normal distribution was motivated by certain requirements which should hold for the mutation operator. In this section, we will clarify why it is useful to use an arbitrary covariance matrix, as in Eq. 2.11, for adaptation.

Any differentiable function $f : \mathbb{R}^n \to \mathbb{R}$ can be approximated by a Taylor series expansion in the vicinity of a position[8] $\tilde{\mathbf{x}} \in \mathbb{R}^n$. Cutting off the Taylor series after the quadratic term, the following approximation is obtained:

$$f(\mathbf{x}) \approx f(\tilde{\mathbf{x}}) + (\mathbf{x} - \tilde{\mathbf{x}})^T \nabla f(\tilde{\mathbf{x}}) + \frac{1}{2}(\mathbf{x} - \tilde{\mathbf{x}})^T \nabla^2 f(\tilde{\mathbf{x}})(\mathbf{x} - \tilde{\mathbf{x}}) \qquad (2.12)$$

Here, $\nabla f(\tilde{\mathbf{x}})$ denotes the gradient, and $\nabla^2 f(\tilde{\mathbf{x}})$ is the symmetric, positive definite *Hessian*, denoted by $\mathbf{H}$ in the following. For a quadratic function f, the Taylor series expansion is exact, and $\mathbf{H}$ contains information about the shape of the isolines of f. In general, these are ellipsoids, as shown in the rightmost part of Fig. 2.1. Hansen describes the relationship between the Hessian $\mathbf{H}$ and the covariance matrix $\mathbf{C}$ of a distribution $N(\mathbf{0}, \mathbf{C})$ informally [29]. It is argued that using $\mathbf{C} = \mathbf{H}^{-1}$ for optimizing a quadratic function is equivalent to using $\mathbf{C} = \mathbf{I}$ for optimizing an isotropic function, such as the sphere function $f(\mathbf{x}) = \frac{1}{2}\mathbf{x}\mathbf{x}^T$.

In other words: Adapting an arbitrary covariance matrix simplifies the optimization by transforming the objective function into an isotropic function. A more formal description of this topic can be found in Rudolph's work, e.g., in the section *Advanced Adaptation Techniques in $\mathbb{R}^n$* in [58], and also in [55].

2.2 Algorithms

This section contains descriptions of the key variants of evolution strategies in chronological order of their publication. On a high level, we differentiate between the two main Sects. 2.2.1 and 2.2.2, with the first one corresponding with the time frame 1964 until 1996.

---

[8] See Sect. 6.2.2.3 in [17].

This first Sect. 2.2.1 describes five main algorithms, namely, the (1+1)-ES as the historically first version of an evolution strategy and the (μ, λ)-MSC-ES (in [58] also called CORR-ES) as the first evolution strategy which adapts an arbitrary covariance matrix (see Sect. 2.1.3 for an explanation). The first derandomized algorithm variants, DR1, DR2, and DR3, complete this selection of older variants of evolution strategies. Their choice is motivated by the fact that they are derandomization steps towards the CMA-ES (see also [63]).

The second main Sect. 2.2.2 describes modern evolution strategies, a term which is used in this book to denote the CMA-ES and algorithms based on it. This distinction might seem somewhat arbitrary, but in fact the development of the CMA-ES defined a turning point in the history of evolution strategies, for two main reasons: First, the CMA-ES is the first algorithm which adapts a covariance matrix in a completely derandomized way. Second, the CMA-ES is seen by many authors as the state of the art in evolution strategies (e.g., [6, 13, 15, 26, 35, 58, 63], and [66]).

2.2.1   From the (1+1)-ES to the CMA-ES

2.2.1.1   (1+1)-ES

The foundation of the first evolution strategy was laid in the 1960s at the Technical University of Berlin by three students, namely Hans-Paul Schwefel, Ingo Rechenberg, and Peter Bienert. As described in [8] or [58], standard methods for solving black-box optimization problems, such as gradient-based methods (see [44]), were not able to deliver satisfactory solution quality for certain optimization problems in engineering applications. Inspired by lectures about biological evolution, they aimed at developing a solution method based on principles of variation and selection. In its first version, a very simple evolution loop without any endogenous parameters was used [59]. This algorithm generates a single offspring $\mathbf{x}' = \mathbf{x} + (N_1(0, \sigma), \dots, N_n(0, \sigma))^T = \mathbf{x} + \sigma \cdot N(\mathbf{0}, \mathbf{I})$ from a single parent individual $\mathbf{x} \in \mathbb{R}^n$. If the offspring performs better than its parent (in terms of fitness), it becomes the new parent. Otherwise, the parent remains. The standard deviation σ of the normal distribution was a fixed scalar value.

According to [53], by pure luck the value of σ was chosen in a way that made this first approach towards a (1+1)-ES successful. Only later on, the necessary step size adaptation was added to the algorithm [52]. Based on two fitness functions, the so-called corridor model[9] and the so-called sphere model,[10] a theoretical result

---

[9]The rectangular corridor model according to [8]: $f_1(\mathbf{x}) = c_0 + c_1 \cdot x_1$ if the constraints $g_j(\mathbf{x})$: $x_j \leq b$ with $b \in \mathbb{R}^+$ for $j \in \{2, \dots, n\}$ are fulfilled, $f_1(\mathbf{x}) = \infty$ otherwise.
[10]The sphere model according to [8]: $f_2(\mathbf{x}) = c_0 + c_1 \cdot \sum_n^{i=1} (x_i - x_i^*)^2$.

Algorithm 2.3 (1+1)-ES

$P_0 \leftarrow \{\mathbf{x}\}$
$\phi \leftarrow f(\mathbf{x})$
$p_S \leftarrow 0$
initialize archive A for storing successful mutations
$t \leftarrow 0$
repeat
   $t \leftarrow t + 1$
   $\mathbf{x}' \leftarrow \mathbf{x} + \sigma \cdot \mathbf{N}(0, \mathbf{I})$
   $\phi' \leftarrow f(\mathbf{x}')$
   if $\phi' < \phi$ **then**
      $\mathbf{x} \leftarrow \mathbf{x}'$
      $\phi \leftarrow \phi'$
      store success in A
   else
      store failure in A
   end if
   $P_t \leftarrow \{\mathbf{x}\}$
   if $t \mod n = 0$ **then**
      get #*successes* and #*failures* from at most $10n$ entries in A
      $p_S = \frac{\#successes}{\#successes + \#failures}$
      $\sigma' \leftarrow \begin{cases} \sigma \cdot c & \text{if } p_S < 1/5 \\ \sigma/c & \text{if } p_S > 1/5 \\ \sigma & \text{if } p_S = 1/5 \end{cases}$
   end if
   $\sigma \leftarrow \sigma'$
until termination criterion fulfilled

---

was derived for introducing step size adaptation: Maximum convergence velocity (i.e., speed of progress of the optimization) is achieved when about 1/5 of all mutations are successful, i.e., improvements over their parent.[11] This insight led to the development of the so-called 1/5-success rule for step size adaptation. If about 1/5 of all mutations are successful, the step size is optimal and no adaptation is required. If the success rate falls below 1/5, the step size needs to be reduced. If it grows above 1/5, the step size needs to be increased. To obtain the new step size $\sigma' = \sigma \cdot c^{\{-1,1\}}$, the previous σ is decreased or increased, respectively, by multiplication or division by $0.817 \leq c \leq 1$. The recommended value of $c = 0.817$ was derived by Schwefel according to theoretical arguments about step size adaptation speed [61]. The step size adaptation according to the above rule is applied each n iterations of the algorithm, and the success rate p_S is measured over a sliding window of the last $10 \cdot n$ mutations [8]. The pseudocode of the (1+1)-ES according to [8] is shown in Algorithm 2.3.

---

[11] The exact values are 0.184 and 0.2025 for the corridor and sphere models, respectively [8].

2.2.1.2 (μ, λ)-MSC-ES

The (μ, λ)-MSC-ES[12] was the very first evolution strategy capable of adapting an arbitrary covariance matrix. The algorithm was developed by Schwefel [62] and is also called (μ, λ)-CORR-ES [58]. In this strategy, the covariance matrix is obtained as a product of $n(n-1)/2$ rotation matrices, where a single rotation matrix R_{ij} for a rotation angle α between axis i and axis j, with $i, j \in \{1, \ldots, n\}$ and $i \neq j$, is given by an identity matrix, extended by the entries $R(i, i) = R(j, j) = \cos \alpha_{ij}$ and $R(i, j) = -R(j, i) = -\sin \alpha_{ij}$.

Indeed, this method is able to generate arbitrary correlated mutations, as proven by Rudolph [55]. In the framework of the (μ, λ)-MSC-ES, endogenous strategy parameters are modified by means of the so-called self-adaptation principle. For self-adaptation, an individual consists not only of the decision parameters $\mathbf{x}$, but also contains an additional vector $\sigma \in \mathbb{R}_+^n$ of step sizes and a vector $\alpha \in (-\pi, \pi]^{n(n-1)/2}$ of rotation angles. The underlying idea of mutative step size adaptation is based on the assumption of individuals with good settings of strategy parameters to generate good offspring, such that the good strategy parameters survive selection. Recombination of decision parameters and endogenous strategy parameters is performed through global intermediary recombination, i.e., by averaging all of the μ parents. Concerning the exogenous strategy parameters, the local and global learning rates τ and τ' need to be set. Following [8], after Schwefel [61], the settings $\tau = \frac{1}{\sqrt{2\sqrt{n}}}$ and $\tau' = \frac{1}{2\sqrt{n}}$ are recommended, depending only on the problem dimensionality n. Pseudocode of the (μ, λ)-MSC-ES is provided in Algorithm 2.4. Concerning the population sizes, we are using $\mu = 15$ and $\lambda = 7 \cdot \mu = 105$ throughout this book, close to the recommendations in [63].

2.2.1.3   DR1

The (μ, λ)-MSC-ES as described in the previous section is based on mutative self-adaptation for step sizes $\delta \in \mathbb{R}_+^n$. However, as Ostermeier et al. [47] claim, self-adaptation of individual step sizes is not possible in the case of small population sizes, and they identify two key reasons: First, a successful mutation of the decision parameters is not necessarily caused by a good step size, but can also be due to an advantageous instantiation of the normally distributed random vector (i.e., a lucky sample). Second, there is a conflict between the goals of maintaining a large variance of step sizes within one generation and avoiding too large fluctuations of step sizes between successive generations. The first derandomized evolution strategy, abbreviated DR1,[13] solves the first problem by using the length of the most successful mutation step within one generation (i.e., the one that yielded the best

---

[12]MSC is an abbreviation of *mutative self-adaptation of covariances*.

[13]In the original publication it is called $(1, \lambda)$-ES with *derandomized mutative step size*.

Algorithm 2.4 (μ, λ)-MSC-ES

initialize population
$P^{(0)} \leftarrow \{(\mathbf{x}_1, \sigma_1, \alpha_1), \ldots, (\mathbf{x}_\mu, \sigma_\mu, \alpha_\mu)\}$
$t \leftarrow 0$
repeat
 $t \leftarrow t + 1$
 // recombination
 $\bar{x} \leftarrow \frac{1}{\mu} \sum_{i=1}^{\mu} \mathbf{x}_i$
 $\bar{\sigma} \leftarrow \frac{1}{\mu} \sum_{i=1}^{\mu} \sigma_i$
 $\bar{\alpha} \leftarrow \frac{1}{\mu} \sum_{i=1}^{\mu} \alpha_i$
 for $i = 1 \rightarrow \lambda$ **do**
  // mutation
  $\eta \leftarrow \tau' \cdot N(0, 1)$
  $\sigma_i \leftarrow \bar{\sigma} \cdot \exp\left(\eta + \tau \cdot N(\mathbf{0}, \mathbf{I})\right)$
  $\alpha_i \leftarrow \bar{\alpha} + \beta \cdot N(\mathbf{0}, \mathbf{I})$
  $\mathbf{C} \leftarrow \prod_{i=1}^{n-1} \prod_{j=i+1}^{n} R_{ij}$
  $\mathbf{x}_i \leftarrow \bar{\mathbf{x}} + \mathbf{C} \cdot \sigma_i \cdot N(\mathbf{0}, \mathbf{I})$
  // evaluation
  $\phi_i \leftarrow f(\mathbf{x}_i)$
 end for
 // selection
 $P^{(t)}$ are the μ best $(\mathbf{x}_i, \sigma_i, \alpha_i)$ from $1 \le i \le \lambda$
until termination criterion fulfilled

offspring) for controlling step size adaptation [47]. The second problem is solved by using a factor $\xi \in \{\frac{5}{7}, \frac{7}{5}\}$ to provide sufficient variance of step sizes within one generation, and to dampen[14] this factor by applying an exponent β with $0 < \beta < 1$ for step size adaptation, to reduce undesired fluctuations [47]. An offspring $\mathbf{x}'$ of a parent $\mathbf{x}$ is then generated as follows:

$$\mathbf{x}' = \mathbf{x} + \xi \cdot \delta \otimes \mathbf{z} \text{ where } \mathbf{z} = N(\mathbf{0}, \mathbf{I})$$

Adaptation of step sizes δ is based on the most successful $\mathbf{z}$ (i.e., the normally distributed vector sample which generated the best offspring during this generation), which is first transformed as follows:

$$\xi_{\mathbf{z}} = \left(\exp\left(|z_1| - \sqrt{2/\pi}\right), \ldots, \exp\left(|z_n| - \sqrt{2/\pi}\right)\right)^T$$

Combined with the exponents β and $\beta_{scal} \in \mathbb{R}$ for damping the adaptation, as well as ξ and $\xi_{\mathbf{z}}$ of the best mutation, the new step sizes δ' are obtained as follows:

$$\delta' = (\xi)^\beta \cdot (\xi_{\mathbf{z}})^{\beta_{scal}} \otimes \delta$$

---

[14]This way, adapting the step size by a factor ξ requires at least $1/\beta > 1$ generations.

Algorithm 2.5 DR1

---

initialize $\mathbf{x}, \boldsymbol{\delta} \leftarrow (1, \ldots, 1)^T$
$t \leftarrow 0$
repeat
$\quad t \leftarrow t + 1$
$\quad$**for** $i = 1 \rightarrow \lambda$ **do**
$\quad\quad \mathbf{z}_i \leftarrow N(\mathbf{0}, \mathbf{I})$
$\quad\quad \mathbf{x}_i \leftarrow \mathbf{x} + \xi_i \cdot \boldsymbol{\delta} \otimes \mathbf{z}_i$ where $P(\xi_i = \frac{5}{7}) = P(\xi_i = \frac{7}{5}) = \frac{1}{2}$
$\quad\quad \phi_i \leftarrow f(\mathbf{x}_i)$
$\quad$**end for**
$\quad sel \leftarrow i$ with best value of ϕ_i
$\quad \mathbf{x} \leftarrow \mathbf{x}_{sel}$
$\quad \xi_{\mathbf{z}_{sel}} = \left(\exp\left(|z_{sel_1}| - \sqrt{2/\pi}\right), \ldots, \exp\left(|z_{sel_n}| - \sqrt{2/\pi}\right) \right)^T$
$\quad \boldsymbol{\delta} \leftarrow (\xi_{sel})^{\beta} \left(\xi_{\mathbf{z}_{sel}}\right)^{\beta_{scal}} \otimes \boldsymbol{\delta}$
until termination criterion fulfilled

Pseudocode of the DR1 evolution strategy is given in Algorithm 2.5. Concerning the offspring population size λ, a constant setting of $\lambda = 10$, independently of dimensionality n, was used in [47]. The DR1 algorithm is based on a single parent individual ($\mu = 1$), and sometimes also denoted as (1, 10)-DR1-ES. Ostermeier et al.[47] recommends for the exponents β and β_{scal} the following values:

$$\beta = \sqrt{1/n}$$

$$\beta_{scal} = 1/n$$

2.2.1.4   DR2

The DR2 evolution strategy[15] represents the next step of derandomization for evolution strategies [48]. The creation of an offspring by mutation is parameterized by a global step size δ and local step sizes $\boldsymbol{\delta}_{scal} \in \mathbb{R}^n$:

$$\mathbf{x}' = \mathbf{x} + \delta \cdot \boldsymbol{\delta}_{scal} \otimes \mathbf{z} \text{ where } \mathbf{z} = N(\mathbf{0}, \mathbf{I})$$

As in DR1, adaptation of step sizes is based on the most successful $\mathbf{z}$. However, in addition to information about the most successful mutation of the current generation, the most successful mutation steps of previous generations are also taken into account, thereby accumulating information over generations. The accumulation takes place in a vector $\boldsymbol{\zeta} \in \mathbb{R}^n$, using a factor $c \in (0, 1]$ to control the weight of previous generations in contrast to the current one:

$$\boldsymbol{\zeta}' = (1 - c) \cdot \boldsymbol{\zeta} + c \cdot \mathbf{z}_{sel} \tag{2.13}$$

---

[15]In the original paper, the algorithm is called $(1, \lambda)$-ES with *derandomized mutative step size control using accumulated information*.

Algorithm 2.6 DR2

---

initialize $\mathbf{x}, \boldsymbol{\zeta} \leftarrow \mathbf{0}, \delta \leftarrow 1, \boldsymbol{\delta}_{scal} \leftarrow (1, \ldots, 1)^T$
$t \leftarrow 0$
repeat
    $t \leftarrow t + 1$
    for $i = 1 \rightarrow \lambda$ **do**
        $\mathbf{z}_i \leftarrow N(\mathbf{0}, \mathbf{I})$
        $\mathbf{x}_i \leftarrow \mathbf{x} + \delta \cdot \boldsymbol{\delta}_{scal} \otimes \mathbf{z}_i$
        $\phi_i \leftarrow f(\mathbf{x}_i)$
    end for
    $sel \leftarrow i$ with best value of ϕ_i
    $\boldsymbol{\zeta}' \leftarrow (1 - c) \cdot \boldsymbol{\zeta} + c \cdot \mathbf{z}_{sel}$
    $\delta' \leftarrow \delta \cdot \left(\exp \left(\frac{\|\boldsymbol{\zeta}'\|}{\sqrt{n} \cdot \sqrt{\frac{c}{2-c}}} - 1 + \frac{1}{5n} \right) \right)^{\beta}$
    $\boldsymbol{\delta}'_{scal} \leftarrow \boldsymbol{\delta}_{scal} \otimes \left(\frac{|\boldsymbol{\zeta}'|}{\sqrt{\frac{c}{2-c}}} + \frac{7}{20} \right)^{\beta_{scal}}$
    $\mathbf{x} \leftarrow \mathbf{x}_{sel}$
    $\boldsymbol{\zeta} \leftarrow \boldsymbol{\zeta}'$
    $\delta \leftarrow \delta'$
    $\boldsymbol{\delta}_{scal} \leftarrow \boldsymbol{\delta}'_{scal}$
until termination criterion fulfilled

Adaptation of step sizes δ and $\boldsymbol{\delta}_{scal}$ is then based on the updated mutation path $\boldsymbol{\zeta}'$:

$$\delta' = \delta \cdot \left(\exp \left(\frac{\|\boldsymbol{\zeta}'\|}{\sqrt{n} \sqrt{\frac{c}{2-c}}} - 1 + \frac{1}{5n} \right) \right)^{\beta}$$

$$\delta'_{scal_i} = \boldsymbol{\delta}_{scal_i} \cdot \left(\frac{|\boldsymbol{\zeta}'_i|}{\sqrt{\frac{c}{2-c}}} + \frac{7}{20} \right)^{\beta_{scal}} \quad \forall i \in \{1, \ldots, n\}$$

Standard settings for the exponents β and β_{scal} as well as the parameter c are as follows:

$$\beta = \sqrt{1/n}$$
$$\beta_{scal} = 1/n$$
$$c = \sqrt{1/n}$$

The pseudocode of the DR2 evolution strategy is given in Algorithm 2.6.

2.2.1.5 DR3

The DR3 evolution strategy [33], also called $(1, \lambda)$-GSA-ES (*generating set adaptation*), is able to generate mutations according to an arbitrary multivariate normal distribution, corresponding to the adaptation of an arbitrary covariance matrix according to Eq. 2.11. This process is not based on implicitly using a covariance matrix, but on transforming an isotropic random vector $\mathbf{z} = N(\mathbf{0}, \mathbf{I})$ into a correlated random vector $\mathbf{y}$ by multiplication with a matrix[16] $\mathbf{B} = (\mathbf{b}_1, \ldots, \mathbf{b}_m) \in \mathbb{R}^{n \times m}$.

As described in Sect. 2.1.3, this can be interpreted as superposition of multiple line distributions. For the number m of column vectors, $n^2 \leq m \leq 2n^2$ holds, with a smaller value of m providing a faster adaptation and a larger value of m a more accurate adaptation. Like in DR1, for variation of the global step size $\delta \in \mathbb{R}$ a factor $\xi \in \{\frac{2}{3}, \frac{3}{2}\}$ with $P(\xi_i = 2/3) = P(\xi_i = 3/2) = 1/2$ is used. To guarantee an approximately constant length of the column vectors in $\mathbf{B}$, $\mathbf{y}$ is adapted by using a factor c_m. Based on its parents $\mathbf{x}$, an offspring is then created as follows:

$$\mathbf{x}' = \mathbf{x} + \delta \cdot \xi \cdot \mathbf{y} \text{ where } \mathbf{y} = c_m \cdot \mathbf{B} N(\mathbf{0}, \mathbf{I})$$

The adaptation of endogenous strategy parameters is based on the selected $\mathbf{y}_{sel}$ and ξ_{sel}. The column vectors of matrix $\mathbf{B}$ are updated according to:

$$\mathbf{b}'_1 = (1 - c) \cdot \mathbf{b}_1 + c \cdot (c_u \xi_{sel} \mathbf{y}_{sel})$$
$$\mathbf{b}'_{i+1} = \mathbf{b}_i \ \forall i \in \{1, \ldots, m - 1\}$$

Like with the previous versions of derandomized evolution strategies, the global step size δ is adapted based on the selected ξ_{sel}, by using a damping exponent β:

$$\delta' = \delta \cdot (\xi_{sel})^\beta$$

For the exogenous parameters, the standard settings are given in [33] as follows:

$$c = \sqrt{1/n}$$
$$\beta = \sqrt{1/n}$$
$$m = \frac{3}{2}n^2$$
$$c_m = (1/\sqrt{m})(1 + 1/m)$$
$$c_u = \sqrt{(2 - c)/c}$$
$$\lambda = 10$$

The corresponding pseudocode of the DR3 evolution strategy is provided in Algorithm 2.7.

---

[16]The column vectors of the matrix $\mathbf{B}$ form a so-called *generating set*, which motivates the terminology *generating set adaptation*.

Algorithm 2.7 DR3

---

initialize $\mathbf{x}, \delta, \mathbf{B} \leftarrow (\mathbf{0}, N(\mathbf{0}, (1/n)\mathbf{I})) \in \mathbb{R}^{n \times m}$
$t \leftarrow 0$
repeat
    $t \leftarrow t + 1$
    for $i = 1 \rightarrow \lambda$ **do**
        $\mathbf{z}_i \leftarrow N(\mathbf{0}, \mathbf{I})$ where $\mathbf{z}_i \in \mathbb{R}^m$
        $\mathbf{y}_i \leftarrow c_m \cdot \mathbf{B}\mathbf{z}_i$
        $\mathbf{x}_i \leftarrow \mathbf{x} + \delta \cdot \xi_i \cdot \mathbf{y}_i$ where $P(\xi_i = 2/3) = P(\xi_i = 3/2) = 1/2$
        $\phi_i \leftarrow f(\mathbf{x}_i)$
    end for
    $sel \leftarrow i$ with best value of ϕ_i
    $\mathbf{b} \leftarrow (1 - c) \cdot \mathbf{b}_1 + c \cdot (c_u \xi_{sel} \mathbf{y}_{sel})$
    $\delta' \leftarrow \delta \cdot (\xi_{sel})^\beta$
    $\mathbf{B}' \leftarrow (\mathbf{b}, \mathbf{b}_1, \ldots, \mathbf{b}_{m-1})$
    $\mathbf{x} \leftarrow \mathbf{x}_{sel}, \delta \leftarrow \delta'$ and $\mathbf{B} \leftarrow \mathbf{B}'$
until termination criterion fulfilled

2.2.2 Modern Evolution Strategies

2.2.2.1 (μ_W, λ)-CMA-ES

Algorithms DR1, DR2 and DR3, as described in Sect. 2.2.1, are derandomized evolution strategies in the sense of adapting endogenous strategy parameters depending on the selected mutation vector. This has also been called the first level of derandomization [63]. In addition, the second level of derandomization aims at the following goals [63]:

- Increase the probability of generating the same mutation step again.
- Provide a direct control mechanism for the rate of change of strategy parameters.
- Keep the strategy parameters unchanged in case of random selection.

The so-called CMA-ES, as introduced in [31], meets these goals by means of two techniques, namely the *covariance matrix adaptation, CMA* and the *cumulative step size adaptation*, CSA, for adapting a global step size. The description of a CMA-ES as provided in [31] is focused on explaining these two techniques, and recombination in case of $\mu > 1$ is not discussed at all. Therefore, we will discuss the CMA-ES in this section as a (μ_W, λ)-CMA-ES with weighted intermediary recombination, as described in [29] and [32].[17] Using the notation for evolution strategies as introduced in Sect. 2.1.2, the algorithm ought to be denoted more precisely as $(\mu/\mu_W, \lambda)$-CMA-ES, with index W denoting the weighted recombination. However, the simplified notation is motivated by arguing that the notation μ/μ_W suggests two different numbers (μ and μ_W), although it is μ in

---

[17] According to [32], the suggestion to use weighted recombination within the CMA-ES is due to Ingo Rechenberg, based on personal communication in 1998.

both cases. Here, we adopt the simplified notation, and denote the CMA-ES with weighted recombination as (μ_W, λ)-CMA-ES.

Based on a parent $\mathbf{x}$, an offspring $\mathbf{x}'$ is then generated as follows:

$$\mathbf{x}' = \mathbf{x} + \sigma \mathbf{B} \mathbf{D} \mathbf{z} \text{ where } \mathbf{z} = N(\mathbf{0}, \mathbf{I})$$

Matrices $\mathbf{B}$ and $\mathbf{D}$ result from an eigendecomposition of the covariance matrix $\mathbf{C}$ according to Eq. 2.7, and $\sigma \in \mathbb{R}$ denotes the global step size. After generating and evaluating an offspring population of size λ according to this mutation operator, the μ best individuals of the offspring population are selected and undergo weighted intermediary recombination.

Weighted intermediary recombination is a generalization of classical global intermediary recombination. Weighted intermediary recombination is based on using μ weights $w_1 \geq w_2 \geq \ldots \geq w_\mu$ with $\sum_{i=1}^{\mu} w_i = 1$ for generating the new parent $\langle \mathbf{x} \rangle$ and the best mutation step $\langle \mathbf{y} \rangle$ as weighted averages:

$$\langle \mathbf{x} \rangle = \sum_{i=1}^{\mu} w_i \mathbf{x}_{i:\lambda}$$

$$\langle \mathbf{y} \rangle = \sum_{i=1}^{\mu} w_i \mathbf{B} \mathbf{D} \mathbf{z}_{i:\lambda}$$

For adapting the strategy parameters, the so-called *variance effective selection mass* μ_{eff} is required:

$$\mu_{eff} = \left(\sum_{i=1}^{\mu} w_i^2 \right)^{-1}$$

According to [29], $1 \leq \mu_{eff} \leq \mu$ holds, and for identical weights $w_i = \frac{1}{\mu}$ ($\forall i \in \{1, \ldots, \mu\}$): $\mu_{eff} = \mu$. In analogy with Eq. 2.13 for DR2, the strategy parameter adaptation techniques, CMA and CSA, use so-called *evolution paths* for accumulating strategy parameter information across several generations. The (μ_W, λ)-CMA-ES uses two evolution paths, $\mathbf{p}_c$ for the adaptation of the covariance matrix and $\mathbf{p}_\sigma$ for global step size adaptation. The evolution paths are updated as follows:

$$\mathbf{p}_c' = (1 - c_c) \cdot \mathbf{p}_c + h_\sigma \sqrt{c_c(2 - c_c)\mu_{eff}} \langle \mathbf{y} \rangle$$

$$\mathbf{p}_\sigma' = (1 - c_\sigma) \cdot \mathbf{p}_\sigma + \sqrt{c_\sigma(2 - c_\sigma)\mu_{eff}} \mathbf{B} \mathbf{D}^{-1} \mathbf{B}^T \langle \mathbf{y} \rangle$$

For updating $\mathbf{p}_c$, the function h_σ is used, which is defined according to:

$$h_\sigma = \begin{cases} 1 & \text{if } \frac{\|\mathbf{p}_\sigma\|}{\sqrt{1-(1-c_\sigma)^{2(t+1)}}} < \left(\frac{7}{5} + \frac{2}{n+1}\right) E(\|N(\mathbf{0}, \mathbf{I})\|) \\ 0 & \text{otherwise} \end{cases}$$

The purpose of h_σ is to avoid an update of $\mathbf{p}_c$ to take information of the current generation t into account, when $\|\mathbf{p}_c\|$ becomes too large. The expectation $E(\|N(\mathbf{0}, \mathbf{I})\|)$ of the length of a multivariate, normally distributed vector of dimensionality n, can be approximated (based on the gamma function[18]) as follows:

$$E(\|N(\mathbf{0}, \mathbf{I})\|) = \sqrt{2}\Gamma(\frac{n+1}{2})/\Gamma(\frac{n}{2}) \approx \sqrt{n}\left(1 - \frac{1}{4n} + \frac{1}{21n^2}\right)$$

The covariance matrix adaptation is performed according to the equation below:

$$\mathbf{C}' = (1 - c_l - c_\mu)\mathbf{C} + c_l(\mathbf{p}_c\mathbf{p}_c^T + \delta(h_\sigma)\mathbf{C}) + c_\mu \sum_{i=1}^{\mu} w_i \mathbf{y}_{i:\lambda}\mathbf{y}_{i:\lambda}^T \qquad (2.14)$$

The first term in the summation represents the contribution of the previous covariance matrix. The second term is called the *rank-one-update* and takes the information accumulated in the evolution path $\mathbf{p}_c$ into account. The third term, the so-called *rank-μ-update*, was introduced with the extension of the CMA-ES for population sizes with $\mu > 1$ [46]. The global step size σ is updated according to:

$$\sigma' = \sigma \cdot \exp\left(\frac{c_\sigma}{d_\sigma}\left(\frac{\|\mathbf{p}_\sigma\|}{E(\|N(\mathbf{0}, \mathbf{I})\|)} - 1\right)\right)$$

For the exogenous strategy parameters of the (μ_W, λ)-CMA-ES, the following standard settings are defined in [29]:

$$\lambda = 4 + \lfloor 3 \ln n \rfloor$$

$$\mu = \lfloor \frac{\lambda}{2} \rfloor$$

$$w_i = \frac{\ln(\frac{\lambda+1}{2}) - \ln i}{\sum_{j=1}^{\mu} \ln(\frac{\lambda+1}{2}) - \ln j} \quad \text{for } i \in \{1, \ldots, \mu\}$$

$$c_\sigma = \frac{\mu_{eff} + 2}{n + \mu_{eff} + 5}$$

$$d_\sigma = 1 + 2\max\left(0, \sqrt{\frac{\mu_{eff} - 1}{n + 1}}\right) + c_\sigma$$

$$c_c = \frac{4 + \mu_{eff}/n}{n + 4 + 2\mu_{eff}/n}$$

---

[18] See [17]: $\Gamma(n) = \int_0^\infty x^{n-1} \exp(-x)\, dx$.

Algorithm 2.8 (μ_W, λ)-CMA-ES

---

 initialize $\langle \mathbf{x} \rangle$
 $\mathbf{p}_c \leftarrow \mathbf{0}$
 $\mathbf{p}_\sigma \leftarrow \mathbf{0}$
 $\mathbf{C} \leftarrow \mathbf{I}$
 $t \leftarrow 0$
 repeat
  $t \leftarrow t + 1$
  $\mathbf{B}$ and $\mathbf{D} \leftarrow$ eigendecomposition of $\mathbf{C}$
  for $i = 1 \rightarrow \lambda$ **do**
   $\mathbf{z}_i \leftarrow N(\mathbf{0}, \mathbf{I})$
   $\mathbf{y}_i \leftarrow \mathbf{BDz}_i$
   $\mathbf{x}_i \leftarrow \langle \mathbf{x} \rangle + \sigma \mathbf{y}_k$
   $f_i \leftarrow f(\mathbf{x}_i)$
  end for
  $\langle \mathbf{y} \rangle \leftarrow \sum_{i=1}^{\mu} w_i \mathbf{y}_{i:\lambda}$
  $\langle \mathbf{x} \rangle \leftarrow \langle \mathbf{x} \rangle + \sigma \langle \mathbf{y} \rangle = \sum_{i=1}^{\mu} w_i \mathbf{x}_{i:\lambda}$
  $\mathbf{p}_\sigma \leftarrow (1 - c_\sigma)\mathbf{p}_\sigma + \sqrt{c_\sigma(2 - c_\sigma)\mu_{eff}} \mathbf{BD}^{-1}\mathbf{B}^T \langle \mathbf{y} \rangle$
  $\sigma \leftarrow \sigma \cdot \exp\left(\frac{c_\sigma}{d_\sigma} \left(\frac{\|\mathbf{p}_\sigma\|}{E\|N(0,\mathbf{I})\|} - 1 \right) \right)$
  $\mathbf{p}_c \leftarrow (1 - c_c)\mathbf{p}_c + h_\sigma \sqrt{c_c(2 - c_c)\mu_{eff}} \langle \mathbf{y} \rangle$
  $\mathbf{C} \leftarrow (1 - c_1 - c_\mu)\mathbf{C} + c_1(\mathbf{p}_c\mathbf{p}_c^T + \delta(h_\sigma)\mathbf{C}) + c_\mu \sum_{i=1}^{\mu} w_i \mathbf{y}_{i:\lambda}\mathbf{y}_{i:\lambda}^T$
 until termination criterion fulfilled

$$c_1 = \frac{2}{\left(n + \frac{13}{10}\right)^2 + \mu_{eff}}$$

$$c_\mu = \min\left(1 - c_1, \alpha_\mu \frac{\mu_{eff} - 2 + 1/\mu_{eff}}{(n + 2)^2 + \alpha_\mu \mu_{eff}/2}\right) \text{ with } \alpha_\mu = 2$$

Putting it all together, the pseudocode of the (μ_W, λ)-CMA-ES is given in Algorithm 2.8.

2.2.2.2 LS-CMA-ES

The LS-CMA-ES [6] is a $(1, \lambda)$-ES implementing the idea to adapt the covariance matrix $\mathbf{C}$ based on the inverse Hessian $\mathbf{H}^{-1}$. The Hessian itself is estimated by solving the appropriate *least squares estimation* problem. Based on Theorem 5 in [55], it is known that this requires at least $m \geq \frac{1}{2}\left(n^2 + 3n + 4\right)$ tuples $(\mathbf{x}, f(\mathbf{x}))$. To achieve this, the algorithm saves all tuples $(\mathbf{x}, f(\mathbf{x}))$ in an archive A. Based on the Taylor series expansion (Eq. 2.12), the *least squares estimation* problem is defined through the following minimization task:

$$\min_{\mathbf{g} \in \mathbb{R}^n, \mathbf{H} \in \mathbb{R}^{n \times n}} \sum_{k=1}^{m} \left(f(\mathbf{x}_k) - f(\mathbf{x}_0) - (\mathbf{x}_k - \mathbf{x}_0)^T \mathbf{g} - \frac{1}{2}(\mathbf{x}_k - \mathbf{x}_0)^T \mathbf{H}(\mathbf{x}_k - \mathbf{x}_0) \right)^2$$

(2.15)

The result of minimizing 2.15 provides estimators $\hat{g}$ for the gradient and $\hat{H}$ for the Hessian.

Since the Taylor series expansion up to the quadratic term provides only an approximation of the true fitness landscape at x_0, we are also interested in obtaining an error measure $Q(\hat{g}, \hat{H})$ of the estimate for deciding whether $\hat{H}^{-1}$ can be used for covariance matrix adaptation. The following error measure is used for this purpose:

$$Q(\hat{g}, \hat{H}) = \frac{1}{m} \sum_{k=1}^{m} \left(\frac{f(x_k) - f(x_0) - (x_k - x_0)^T \hat{g} - \frac{1}{2}(x_k - x_0)^T \hat{H}(x_k - x_0)}{f(x_k) - f(x_0) - (x_k - x_0)^T \hat{g}} \right)^2$$

(2.16)

Unfortunately, solving Eq. 2.15 and inverting $\hat{H}$ by means of numerical methods requires algorithms with time complexity $O(n^6)$, so that, especially for large n, an execution of these steps in each generation is not affordable. To solve this problem, the LS-CMA-ES provides two different working modes, denoted LS and CMA, for adapting the covariance matrix.

In mode LS, an approximation of H is performed only each n_{upd} generations.[19] If the error Q falls below a required threshold Q_t, the covariance matrix $C = \frac{1}{2}\hat{H}^{-1}$ is used by the algorithm and remains unchanged until a new update after another n_{upd} generations is performed.

If Q is bigger than the threshold value Q_t, the LS-CMA-ES switches into mode CMA. Before explaining this mode, the creation of an offspring x' from the parent $\langle x \rangle$ is defined below:

$$x' = \langle x \rangle + \sigma d N(0, C) \text{ where } d = \exp(\tau N(0, 1))$$

In addition to the covariance matrix C, a global step size σ is used, which is updated by mutative step size adaptation. If b denotes the index of the best offspring, the global step size is changed according to $\sigma' = \sigma \cdot d_b$. Adapting the covariance matrix C is based on a *rank-one update* (i.e., the second term in Eq. 2.14) by using an evolution path p_c:

$$p_c' = (1 - c_c) \cdot p_c + \frac{\sqrt{(c_c(2 - c_c))}}{\sigma}(x_b - \langle x \rangle)$$

$$C' = (1 - c_{cov}) \cdot C + c_{cov} p_c (p_c)^T$$

The evolution path p_c is also updated when operating in mode LS, to make sure C is updated based on up-to-date information when the algorithm switches into mode CMA.

The pseudocode of the LS-CMA-ES is given in Algorithm 2.9, and the exogenous strategy parameters are set as follows:

---

[19]With the additional condition for A to consist of at least $m = n^2$ tuples.

$$\lambda = 10$$

$$\tau = \frac{1}{\sqrt{n}}$$

$$n_{upd} = 100$$

$$Q_t = 10^{-3}$$

$$c_c = \frac{4}{n + 4}$$

$$c_{cov} = \frac{2}{(n + \sqrt{2})^2}$$

2.2.2.3   LR-CMA-ES

The LR-CMA-ES (*local restart*) extends the (μ_W, λ)-CMA-ES by introducing restarts [4]. The strategy introduces five criteria for identifying stagnation of the optimization process and, in case of stagnation, starts a new run of the (μ_W, λ)-CMA-ES. Each run of the (μ_W, λ)-CMA-ES initializes the starting point of the search and the strategy parameters anew, so that the runs are independent of each other. For defining the termination criteria, the tolerance values $T_x = \sigma 10^{-12}$ and $T_f = 10^{-12}$ are used. Any other exogenous parameters are the same as in the (μ_W, λ)-CMA-ES.

The first termination criterion, called *equalfunvalhist*, is satisfied if either the best fitness values $f(\mathbf{x}_{1:\lambda})$ of the last $\lceil 10 + 30n/\lambda \rceil$ generations are identical or the difference between their maximum and minimum values is smaller than T_x.

The second criterion, *TolX*, is satisfied if the components of the vector $\mathbf{v} = \sigma \mathbf{p}_c$ are all smaller than T_x, i.e., $v_i < T_x \; \forall i \in \{1, \ldots, n\}$.

The third criterion, *noeffectaxis*, takes changes with respect to the main coordinate axes induced by $\mathbf{C}$ into account. These are given by the eigenvectors $\mathbf{u}_i$ and eigenvalues $\gamma_i, i \in \{1, \ldots, n\}$, of $\mathbf{C}$, and they are found (normalized) in the columns of matrix $\mathbf{B}$ and the main diagonal elements of $\mathbf{D}$. The termination criterion does not check all main axes at once, but in generation t it takes the axis $i = t \bmod n$ into account. It is satisfied when $\frac{\sigma}{10} \sqrt{\gamma_i} \mathbf{u}_i \approx 0$.

The fourth criterion, *noeffectcoord*, analyzes changes with respect to the coordinate axes. It is satisfied if $\frac{\sigma}{5} C_{i,i} \approx 0 \; \forall i \in \{1, \ldots, n\}$.

Finally, the criterion *conditioncov* checks whether the condition number of the matrix $\mathbf{C}$, $\mathrm{cond}(\mathbf{C}) = \frac{\max(\{\gamma_1, \ldots, \gamma_n\})}{\min(\{\gamma_1, \ldots, \gamma_n\})}$ exceeds 10^{14}.

The pseudocode of the LR-CMA-ES, as shown in Algorithm 2.10, consists of a simple outer loop managing the restarts of the (μ_W, λ)-CMA-ES. The local termination criteria are exactly the five criteria introduced above for discovering stagnation. In contrast, the global termination criterion is the same as used in previous sections, see Sect. 2.1.2.

Algorithm 2.9 LS-CMA-ES

---

initialize $\langle \mathbf{x} \rangle, \sigma$
$\mathbf{C} \leftarrow \mathbf{I}$
Archive $A \leftarrow \emptyset$
$\mathbf{p}_c \leftarrow \mathbf{0}$
mode $\leftarrow$ LS
$t \leftarrow 0$
repeat
   $t \leftarrow t + 1$
   $\mathbf{B}$ and $\mathbf{D} \leftarrow$ eigendecomposition of $\mathbf{C}$
   for $i = 1 \rightarrow \lambda$ **do**
      $d_i \leftarrow \exp(\tau N(0, 1))$
      $\mathbf{x}_i \leftarrow \langle \mathbf{x} \rangle + \sigma \cdot d_i \mathbf{B}\mathbf{D}N(\mathbf{0}, \mathbf{I})$
      $f_i \leftarrow f(\mathbf{x}_i)$
      $A \leftarrow A \cup \{(\mathbf{x}_i, f_i)\}$
   end for
   $b \leftarrow$ index of best offspring
   $\sigma \leftarrow \sigma \cdot d_b$
   $\mathbf{p}_c \leftarrow (1 - c_c)\mathbf{p}_c + \frac{\sqrt{c_c(2-c_c)}}{\sigma}(\langle \mathbf{x} \rangle - \mathbf{x}_b)$
   if mode = LS **then**
      $\mathbf{C}$ unchanged
   else if mode = CMA **then**
      $\mathbf{C} \leftarrow (1 - c_{cov})\mathbf{C} + c_{cov}\mathbf{p}_c\mathbf{p}_c^T$
   end if
   if t modulo $n_{upd} = 0$ **then**
      Obtain $\hat{\mathbf{g}}$ and $\hat{\mathbf{H}}$ based on the last n^2 tuples of A by solving Equation 2.15 where $\mathbf{x}_0 = \langle \mathbf{x} \rangle$.
      Obtain $Q(\hat{\mathbf{g}}, \hat{\mathbf{H}})$ from Equation 2.16
      if $Q(\hat{\mathbf{g}}, \hat{\mathbf{H}}) < Q_t$ **then**
         mode $\leftarrow$ LS
         $\mathbf{C} \leftarrow \left(\frac{1}{2}\hat{\mathbf{H}}\right)^{-1}$
      else
         mode $\leftarrow$ CMA
      end if
   end if
   $\langle \mathbf{x} \rangle \leftarrow \mathbf{x}_b$
until termination criterion fulfilled

Algorithm 2.10 LR-CMA-ES

---

repeat
   execute (μ_W, λ)-CMA-ES (Algorithm 2.8) using the local termination criteria
until global termination criterion satisfied

2.2.2.4   IPOP-CMA-ES

The IPOP-CMA-ES [5] is an extension of the LR-CMA-ES as described in the previous section. Whenever a run of the (μ_W, λ)-CMA-ES is terminated due to a local termination criterion (as introduced for LR-CMA-ES), the population size is increased by a factor η for the next run of the (μ_W, λ)-CMA-ES. This strategy is

---

Algorithm 2.11 IPOP-CMA-ES

---
repeat
    execute (μ_W, λ)-CMA-ES (Algorithm 2.8) using the local termination criteria
    $\mu \leftarrow \eta \cdot \mu$
    $\lambda \leftarrow \eta \cdot \lambda$
until global termination criterion satisfied

---

motivated by empirical investigations of the behavior of the (μ_W, λ)-CMA-ES with different population sizes for multimodal test functions [30]. As these investigations clarified, the global convergence properties of the algorithm improve with increasing population size. The corresponding pseudocode is given in Algorithm 2.11. When using non-integer values for η, the new number of parents μ and offspring λ are obtained by rounding. For η, the interval $\left[\frac{3}{2}, 5\right]$ is identified as a reasonable range, and the default value $\eta = 2$ is recommended.

2.2.2.5   (1+1)-Cholesky-CMA-ES

The (1+1)-Cholesky-CMA-ES [38] introduces a method for adapting the covariance matrix $\mathbf{C}$ implicitly, without using an eigendecomposition of $\mathbf{C}$. Consequently, the approach reduces the computational complexity within each generation from $O(n^3)$ to $O(n^2)$.

The algorithm is based on the so-called Cholesky decomposition[20] of the covariance matrix, $\mathbf{C} = \mathbf{A}\mathbf{A}^T$. As proven in [38], an update of the Cholesky factors $\mathbf{A}$ is possible without explicit knowledge of the covariance matrix $\mathbf{C}$. The corresponding lemma and theorem are stated here without proof. The lemma states that, for any vector $\mathbf{v} \in \mathbb{R}^n$ and $\varsigma = \frac{1}{\|\mathbf{v}\|^2}\left(\sqrt{1 + \|\mathbf{v}\|^2} - 1\right)$, the following equation holds:

$$\mathbf{I} + \mathbf{v}\mathbf{v}^T = \left(\mathbf{I} + \varsigma \mathbf{v}\mathbf{v}^T\right)\left(\mathbf{I} + \varsigma \mathbf{v}\mathbf{v}^T\right)$$

This lemma is required for the proof of the following theorem:

Theorem 2.2.1. *Let $\mathbf{C} \in \mathbb{R}^n$ be a symmetric, positive definite matrix with Cholesky decomposition $\mathbf{C} = \mathbf{A}\mathbf{A}^T$. Let $\mathbf{C}' = \alpha\mathbf{C} + \beta\mathbf{v}\mathbf{v}^T$ be an update of $\mathbf{C}$ with $\mathbf{v}, \mathbf{z} \in \mathbb{R}^n$, $\mathbf{v} = \mathbf{A}\mathbf{z}$ and $\alpha, \beta \in \mathbb{R}^+$. The updated Cholesky factor $\mathbf{A}'$ of $\mathbf{C}'$ is then given by*

$$\mathbf{A}' = \sqrt{\alpha}\mathbf{A} + \frac{\sqrt{\alpha}}{\|\mathbf{z}\|^2}\left(\sqrt{1 + \frac{\beta}{\alpha}\|\mathbf{z}\|^2} - 1\right)(\mathbf{A}\mathbf{z})\,\mathbf{z}^T.$$

Based on a parent individual $\mathbf{x}$, an offspring $\mathbf{x}'$ is then created according to:

$$\mathbf{x}' = \mathbf{x} + \sigma\mathbf{A}\mathbf{z} \text{ with } \mathbf{z} = N(\mathbf{0}, \mathbf{I})$$

---

[20]Compare Sect. 19.2.1.2 in [17].

Using Theorem 2.2.1, the Cholesky factor $\mathbf{A}$ is adapted as follows:

$$\mathbf{A}' = c_a \mathbf{A} + \frac{c_a}{\|\mathbf{z}\|^2} \left(\sqrt{1 + \frac{(1 - c_a^2)\|\mathbf{z}\|^2}{c_a^2}} - 1 \right) \mathbf{A}\mathbf{z}\mathbf{z}^T,$$

with a constant exogenous strategy parameter c_a. The adaptation above is applied if the value of a measure $\bar{p}_s$ (explained in the following) is smaller than a threshold value p_t.

The adaptation of the global step size δ is in some ways similar to the 1/5-success rule of the (1+1)-ES (see Sect. 2.2.1). If the offspring is better than the parent, $\lambda_s = 1$ in the equation below, otherwise, $\lambda_s = 0$. These success indicators are accumulated across generations by using a learning rate c_p, resulting in an accumulated success rate $\bar{p}_s$:

$$\bar{p}_s = (1 - c_p)\bar{p}_s + c_p \lambda_s$$

Using this measure and its target value p_s^t for the success rate, the global step size σ is updated as follows:

$$\sigma' = \sigma \cdot \exp\left(\frac{1}{d} \left(\bar{p}_s - \frac{p_s^t}{1 - p_s^t}(1 - \bar{p}_s) \right) \right)$$

The pseudocode is given in Algorithm 2.12, and the default settings of the exogenous strategy parameters are:

$$p_s^t = \frac{2}{11}$$

$$p_t = \frac{11}{25}$$

$$c_a = \sqrt{1 - \frac{2}{n^2 + 6}}$$

$$c_p = \frac{1}{12}$$

$$d = 1 + \frac{1}{n}$$

2.2.2.6 Active-CMA-ES

The (μ_W, λ)-CMA-ES uses weighted recombination of the μ best offspring to generate a new point in the search space. As shown by Rudolph [57], the convergence velocity of an evolution strategy can be further increased by also taking

---

Algorithm 2.12 (1+1)-Cholesky-CMA-ES

---

   initialize $\mathbf{x}, \sigma$
   $\mathbf{A} \leftarrow \mathbf{I}$
   $\bar{p}_s \leftarrow p_s^t$
   repeat
      $\mathbf{z} \leftarrow N(\mathbf{0}, \mathbf{I})$
      $\mathbf{x}' \leftarrow \mathbf{x} + \sigma \mathbf{A} \mathbf{z}$
      if $f(\mathbf{x}') \leq f(\mathbf{x})$ **then**
         $\lambda_s \leftarrow 1$
      else
         $\lambda_s \leftarrow 0$
      end if
      $\bar{p}_s \leftarrow (1 - c_p)\bar{p}_s + c_p \lambda_s$
      $\sigma \leftarrow \sigma \cdot \exp\left(\frac{1}{d}\left(\bar{p}_s - \frac{p_s^t}{1 - p_s^t}(1 - \bar{p}_s)\right)\right)$
      if $f(\mathbf{x}') \leq f(\mathbf{x})$ **then**
         $\mathbf{x} \leftarrow \mathbf{x}'$
         if $\bar{p}_s \leq p_t$ **then**
$$\mathbf{A} \leftarrow c_a \mathbf{A} + \frac{c_a}{\|\mathbf{z}\|^2}\left(\sqrt{1 + \frac{(1 - c_a^2)\|\mathbf{z}\|^2}{c_a^2}} - 1\right)\mathbf{A}\mathbf{z}\mathbf{z}^T$$
         end if
      end if
   until termination criterion satisfied

---

the worst offspring into account for recombination, however, with negative weights. The Active-CMA-ES [40] is based on this idea,[21] however, it is not used during the process of recombination,[22] but exclusively for adapting the covariance matrix. Therefore, the corresponding extension of the (μ_W, λ)-CMA-ES mainly consists of changing the covariance matrix adaptation method, modifying Eq. 2.14 of the (μ_W, λ)-CMA-ES within the Active-CMA-ES into:

$$\mathbf{C}' = \mathbf{C} \leftarrow (1 - c_c)\mathbf{C} + c_c \mathbf{p}_c \mathbf{p}_c^T + \beta \mathbf{Z} \text{ where}$$

$$\mathbf{Z} = \mathbf{BD}\left(\frac{1}{\mu}\sum_{k=1}^{\mu} \mathbf{z}_{k:\lambda}\mathbf{z}_{k:\lambda}^T - \frac{1}{\mu}\sum_{k=\lambda-\mu+1}^{\lambda} \mathbf{z}_{k:\lambda}\mathbf{z}_{k:\lambda}^T\right)(\mathbf{BD})^T$$

In addition, the exogenous parameter c_c is now modified to $c_c = \frac{2}{(n+\sqrt{2})^2}$. The parameter β has been tuned by means of an empirical investigation, which is described in detail in [39]. Its setting of $\beta = \frac{4\mu-2}{(n+12)^2+4\mu}$ reflects a compromise between the conflicting goals of achieving a large convergence velocity on the one

---

[21] The term *active* is motivated by the fact that specifically the bad offspring individuals play an active role, although they would normally not be taken into account after selection has been applied.
[22] This is explicitly avoided due to the occurrence of numerical instabilities for certain objective functions; see [40].

Algorithm 2.13 Active-CMA-ES

---

initialize $\langle \mathbf{x} \rangle$
$\mathbf{p}_c \leftarrow \mathbf{0}$
$\mathbf{p}_\sigma \leftarrow \mathbf{0}$
$\mathbf{C} \leftarrow \mathbf{I}$
$t \leftarrow 0$
repeat
    $t \leftarrow t + 1$
    $\mathbf{B}$ and $\mathbf{D} \leftarrow$ from eigendecomposition of $\mathbf{C}$
    for $i = 1 \rightarrow \lambda$ **do**
        $\mathbf{z}_i \leftarrow N(\mathbf{0}, \mathbf{I})$
        $\mathbf{y}_i \leftarrow \mathbf{BD}\mathbf{z}_i$
        $\mathbf{x}_i \leftarrow \langle \mathbf{x} \rangle + \sigma \mathbf{y}_k$
        $f_i \leftarrow f(\mathbf{x}_i)$
    end for
    $\langle \mathbf{y} \rangle \leftarrow \sum_{i=1}^{\mu} w_i \mathbf{y}_{i:\lambda}$
    $\langle \mathbf{x} \rangle \leftarrow \langle \mathbf{x} \rangle + \sigma \langle \mathbf{y} \rangle = \sum_{i=1}^{\mu} w_i \mathbf{x}_{i:\lambda}$
    $\mathbf{p}_\sigma \leftarrow (1 - c_\sigma)\mathbf{p}_\sigma + \sqrt{c_\sigma(2 - c_\sigma)\mu_{\mathit{eff}}}\mathbf{BD}^{-1}\mathbf{B}^T \langle \mathbf{y} \rangle$
    $\sigma \leftarrow \sigma \cdot \exp\left(\frac{c_\sigma}{d_\sigma} \left(\frac{\|\mathbf{p}_\sigma\|}{E\|N(\mathbf{0},\mathbf{I})\|} - 1 \right) \right)$
    $\mathbf{p}_c \leftarrow (1 - c_c)\mathbf{p}_c + h_\sigma \sqrt{c_c(2 - c_c)\mu_{\mathit{eff}}} \langle \mathbf{y} \rangle$
    $\mathbf{Z} \leftarrow \mathbf{BD} \left(\frac{1}{\mu} \sum_{k=1}^{\mu} \mathbf{z}_{k:\lambda}\mathbf{z}_{k:\lambda}^T - \frac{1}{\mu} \sum_{k=\lambda-\mu+1}^{\lambda} \mathbf{z}_{k:\lambda}\mathbf{z}_{k:\lambda}^T \right) (\mathbf{BD})^T$
    $\mathbf{C} \leftarrow (1 - c_c)\mathbf{C} + c_c\mathbf{p}_c\mathbf{p}_c^T + \beta\mathbf{Z}$
until termination criterion satisfied

---

hand and ensuring that $\mathbf{C}$ remains positive definite, to drive the evolution strategy into a robust working regime. The pseudocode is provided in Algorithm 2.13, and the default settings of the exogenous strategy parameters are, except for c_c and β, identical to those used in the (μ_W, λ)-CMA-ES.

2.2.2.7 (μ,λ)-CMSA-ES

The (μ,λ)-CMSA-ES [13], more precisely denoted the $(\mu/\mu_I, \lambda)$-CMA-σ-SA-ES, reintroduces self-adaptation of the global step size σ, just like in the (μ, λ)-MSC-ES, into the algorithm. This approach is motivated by the fact that reintroducing self-adaptation decreases the number of exegenous strategy parameters to two,[23] consequently providing a simplification of the (μ_W, λ)-CMA-ES, which requires five exogenous strategy parameters. Offspring individuals $\mathbf{x}_i$ and their step sizes σ_i, $i \in \{1, \ldots, \lambda\}$, are created based on the parent $\mathbf{x}$, the global step size σ, and the matrices $\mathbf{B}$ and $\mathbf{D}$ (from an eigendecomposition of the covariance matrix $\mathbf{C}$), as follows:

---

[23]Population sizes μ and λ are not counted.

$$\sigma_i = \sigma \cdot \exp\left(\tau N(0,1)\right)$$

$$\mathbf{s}_i = \mathbf{BD}N(\mathbf{0}, \mathbf{I})$$

$$\mathbf{z}_i = \sigma_i \cdot \mathbf{s}_i$$

$$\mathbf{x}_i = \mathbf{x} + \mathbf{z}_i$$

Recombination is based on identical weights $1/\mu$, resulting in averaging the μ best offspring. It is applied to the vectors $\mathbf{z}_{i:\lambda}$, $\mathbf{s}_{i:\lambda}$, and step sizes $\sigma_{i:\lambda}$, for $i \in \{1,\ldots,\mu\}$, and results in the vectors $\langle \mathbf{z} \rangle$, $\langle \mathbf{s} \rangle$ and the new global step size σ. The new parent $\mathbf{x}'$ is then obtained as $\mathbf{x}' = \mathbf{x} + \langle \mathbf{z} \rangle$. Vector $\langle \mathbf{s} \rangle$ is required for adapting the covariance matrix $\mathbf{C}$, and its update uses the learning rate τ_C by proceeding as follows:

$$\mathbf{C}' = \left(1 - \frac{1}{\tau_C}\right)\mathbf{C} + \frac{1}{\tau_C}\langle \mathbf{s} \rangle \langle \mathbf{s} \rangle^T \tag{2.17}$$

The default settings of the exogenous strategy parameters are:

$$\mu = \max\left(\left\lfloor \frac{n}{10} \right\rfloor, 2\right)$$

$$\lambda = 4\mu$$

$$\tau = \frac{1}{\sqrt{2n}}$$

$$\tau_C = 1 + \frac{n(n+1)}{2\mu}$$

The pseudocode of the corresponding (μ,λ)-CMSA-ES is given in Algorithm 2.14.

2.2.2.8 sep-CMA-ES

The sep-CMA-ES [54] is a variation of the (μ_W, λ)-CMA-ES which reduces space and time complexity to reach $O(n)$, i.e., linear in n. This is achieved by using, instead of an arbitrary covariance matrix, just a diagonal matrix $\mathbf{D}$ as in Eq. 2.10. Consequently, this kind of evolution strategy is not able anymore to generate correlated mutations, in return for the advantage of saving the computationally intensive eigendecomposition of the covariance matrix $\mathbf{C}$. $\mathbf{D}$ can then be obtained from $\mathbf{C}$ by taking the square roots of the main diagonal elements of $\mathbf{C}$. The covariance matrix is adapted according to the following update rule:

$$\mathbf{C}' = (1 - c_{cov})\mathbf{C} + \frac{1}{\mu_{eff}}c_{cov}\mathbf{p}_c(\mathbf{p}_c)^T + c_{cov}\left(1 - \frac{1}{\mu_{eff}}\right)\sum_{i=1}^{\mu} w_i \mathbf{Dz}_{i:\lambda}(\mathbf{Dz}_{i:\lambda})^T$$

Algorithm 2.14 (μ,λ)-CMSA-ES

initialize $\mathbf{x}$, σ
$\mathbf{C} \leftarrow \mathbf{I}$
$\langle \sigma \rangle \leftarrow \sigma$
repeat
   $\mathbf{B}$ and $\mathbf{D} \leftarrow$ from eigendecomposition of $\mathbf{C}$
   for $i = 1 \rightarrow \lambda$ **do**
      $\sigma_i \leftarrow \langle \sigma \rangle \exp \tau N(0,1)$
      $\mathbf{s}_i \leftarrow \mathbf{B}\mathbf{D}N(\mathbf{0},\mathbf{I})$
      $\mathbf{z}_i \leftarrow \sigma_i \cdot \mathbf{s}_i$
      $\mathbf{y}_i \leftarrow \mathbf{x} + \mathbf{z}_i$
      $f_i \leftarrow f(\mathbf{y}_i)$
   end for
   $\langle \mathbf{z} \rangle \leftarrow$ average of the best μ $\mathbf{z}_i, i \in \{1,\ldots,\lambda\}$
   $\langle \mathbf{s} \rangle \leftarrow$ average of the best μ $\mathbf{s}_i, i \in \{1,\ldots,\lambda\}$
   $\langle \sigma \rangle \leftarrow$ average of the best μ $\sigma_i, i \in \{1,\ldots,\lambda\}$
   $\mathbf{x} \leftarrow \mathbf{x} + \langle \mathbf{z} \rangle$
   $\mathbf{C} \leftarrow \left(1 - \frac{1}{\tau_C}\right) \mathbf{C} + \frac{1}{\tau_C} \langle \mathbf{s}\mathbf{s}^T \rangle$
until termination criterion satisfied

Due to the reduced complexity of the covariance matrix, the learning rate c_{cov} can be increased to accelerate the adaptation process. The learning rate c_{cov} is then set as follows:

$$c_{cov} = \frac{n+2}{3} \left(\frac{1}{\mu_{eff}} \frac{2}{(n+\sqrt{2})^2} + (1 - \frac{1}{\mu_{eff}}) \min \left(1, \frac{2\mu_{eff} - 1}{(n+2)^2 + \mu_{eff}} \right) \right)$$

All other settings of the sep-CMA-ES are identical to those used within the (μ_W, λ)-CMA-ES. The resulting pseudocode of the sep-CMA-ES is shown in Algorithm 2.15.

2.2.2.9 $(1 \overset{+}{,} \lambda^s_m)$-ES

The $(1 \overset{+}{,} \lambda^s_m)$-ES [16] introduces the two new concepts of *mirrored sampling* and *sequential selection*. These two mutually independent concepts change the algorithmic processes of offspring creation and their selection, and thus they do not establish a complete evolution strategy. The concept of *mirrored sampling* can be used within a $(1 + \lambda)$-ES as well as a $(1, \lambda)$-ES. The application of *sequential selection* is only possible in the case of a plus-strategy, explaining also the use of the notation $\overset{+}{,}$. Furthermore, the indices s and m of λ represent the algorithmic concepts of *sequential selection* (s) and *mirrored sampling* (m), respectively.

The idea of *mirrored sampling* is to generate part of the offspring in a derandom-ized way by generating for a mutation vector $\mathbf{z}$ not only the offspring $\mathbf{x} + \mathbf{z}$, but also

---

Algorithm 2.15 sep-CMA-ES

---

initialize $\langle \mathbf{x} \rangle$
$\mathbf{C} \leftarrow \mathbf{I}$
$\mathbf{D} \leftarrow \mathbf{I}$
$\mathbf{p}_\sigma \leftarrow \mathbf{0}$
$\mathbf{p}_c \leftarrow \mathbf{0}$
$t \leftarrow 0$
repeat
    $t \leftarrow t + 1$
    for $i = 1 \rightarrow \lambda$ **do**
        $\mathbf{z}_i \leftarrow N(\mathbf{0}, \mathbf{I})$
        $\mathbf{x}_i \leftarrow \langle \mathbf{x} \rangle + \sigma \mathbf{D}\mathbf{z}_i$
    end for
    $\langle \mathbf{x} \rangle \leftarrow \sum_{i=1}^{\mu} w_i \mathbf{x}_{i:\lambda}$
    $\langle \mathbf{z} \rangle \leftarrow \sum_{i=1}^{\mu} w_i \mathbf{z}_{i:\lambda}$
    $\mathbf{p}_\sigma \leftarrow (1 - c_\sigma)\mathbf{p}_\sigma + \sqrt{c_\sigma(2 - c_\sigma)}\sqrt{\mu_{eff}}\langle \mathbf{z} \rangle$
    if $\frac{\|\mathbf{p}_\sigma\|}{\sqrt{1 - (1 - c_\sigma)^{2t}}} < \left(\frac{7}{5} + \frac{2}{n+1} \right) E(\|N(\mathbf{0}, \mathbf{I})\|)$ **then**
        $H_\sigma \leftarrow 1$
    else
        $H_\sigma \leftarrow 0$
    end if
    $\mathbf{p}_c \leftarrow (1 - c_c)\mathbf{p}_c + H_\sigma \sqrt{c_c(2 - c_c)}\sqrt{\mu_{eff}}\mathbf{D}\langle \mathbf{z} \rangle$
    $\mathbf{C} \leftarrow (1 - c_{cov})\mathbf{C} + \frac{c_{cov}}{\mu_{eff}}\mathbf{p}_c \mathbf{p}_c^T + c_c \left(1 - \frac{1}{\mu_{eff}} \right) \sum_{i=1}^{\mu} w_i \mathbf{D}\mathbf{z}_{i:\lambda} (\mathbf{D}\mathbf{z}_{i:\lambda})^T$
    $\sigma \leftarrow \sigma \exp \left(\frac{c_\sigma}{d_\sigma} \left(\frac{\|\mathbf{p}_\sigma\|}{E(\|N(\mathbf{0}, \mathbf{I})\|)} - 1 \right) \right)$
    $\mathbf{D} = \text{diag} \left(\sqrt{C_{1,1}}, \ldots, \sqrt{C_{n,n}} \right)$
until termination criterion satisfied

---

the additional offspring $\mathbf{x} - \mathbf{z}$. These two offspring are obviously symmetrical[24] with respect to $\mathbf{x}$. As a potential application, mentioned in [3], *mirrored sampling* can increase the robustness of the *Evolutionary Gradient Search* algorithm and increase convergence velocity in the sphere model. Theoretical convergence rates for variants of the $(1 \overset{+}{,} \lambda_m^s)$-ES have been derived; see [16] for the corresponding results.

Sequential selection can be used to reduce the number of function evaluations. It is applied within a $(1 + \lambda)$-ES by sequentially executing the steps mutation and evaluation for single offspring individuals, rather than generating all λ offspring first and then evaluating their fitness. In *sequential selection*, as soon as an offspring has a better fitness than the parent, the offspring can replace the parent, and no more offspring need to be generated and evaluated. In this way, up to $\lambda - 1$ function evaluations can potentially be saved at each generation.

The two concepts can be used independently of each other, or in combination. As explained before, the $(1 \overset{+}{,} \lambda_m^s)$-ES does not constitute a complete evolution strategy, but rather a method for generating the parent $\langle \mathbf{x} \rangle'$ for the next generation based on the previous parent $\langle \mathbf{x} \rangle$ and a method *mutationOffset*, which generates a

---

[24]Instead of the term symmetrical, this is called *mirrored* in the context of this strategy.

Algorithm 2.16 $(1 \overset{+}{,} \lambda_m^s)$-ES

Input: search point $\langle \mathbf{x} \rangle$ and a method *mutationOffset*
Output: new search point $\langle \mathbf{x} \rangle'$

$i \leftarrow 0$
$j \leftarrow 0$
while $i < \lambda$ **do**
  $i \leftarrow i + 1$
  $j \leftarrow j + 1$
  if *(mirrored sampling)* $\wedge$ $(j \bmod 2 = 0)$ **then**
    $\mathbf{x}_i \leftarrow \langle \mathbf{x} \rangle - \mathbf{z}_i$
  else
    $\mathbf{z}_i \leftarrow mutationOffset()$
    $\mathbf{x}_i \leftarrow \langle \mathbf{x} \rangle + \mathbf{z}_i$
  end if
  if *(sequential selection)* $\wedge$ $(f(\mathbf{x}_i) < f(\langle \mathbf{x} \rangle))$ **then**
    $j \leftarrow 0$
    break
  end if
end while
$\langle \mathbf{x} \rangle' \leftarrow \operatorname{argmin}(\{f(\mathbf{x}_1), \ldots, f(\mathbf{x}_i)\})$

---

mutation step and is determined by the underlying evolution strategy. The approach is summarized in pseudocode in Algorithm 2.16.

2.2.2.10 xNES

The xNES algorithm *(exponential natural evolution strategies)* [26] is a $(1, \lambda)$-ES which adapts its endogenous strategy parameters by using the so-called *natural gradient* (see [1]). The idea was implemented for the first time in the context of NES *(natural evolution strategies)* [71] and was then developed further by introducing the eNES *(efficient natural evolution strategies)*[25] [66].

In the following, the underlying ideas of the xNES are briefly summarized, without giving detailed descriptions of the underlying concepts, such as, e.g., the *Fisher information matrix*. These fundamentals can be found in the original work of Glasmachers et al. and the corresponding references, see [26].

This family of evolution strategy algorithms also relies on the multivariate normal distribution $N(\langle \mathbf{x} \rangle, \mathbf{C})$ for generating correlated mutations of the current search point $\langle \mathbf{x} \rangle$. Similar to the $(1 + 1)$-Cholesky-CMA-ES (see Sect. 2.2.2.5), rather than working with the covariance matrix $\mathbf{C}$ explicitly, a Cholesky factor $\mathbf{A}$ with $\mathbf{C} = \mathbf{A}\mathbf{A}^T$ is used. The current search point and the covariance matrix are combined to form the tuple $\theta = (\langle \mathbf{x} \rangle, \mathbf{C})$, representing the quantities subject to adaptation within an xNES. Rewriting the probability density function of a normal distribution

---

[25] In [26] the eNES are called *exact natural evolution strategies*.

as a function of the current search point $\langle \mathbf{x} \rangle$ and the Cholesky factor $\mathbf{A}$, its probability density $N(\langle \mathbf{x} \rangle, \mathbf{C})$ turns into:

$$p(\mathbf{x}|\theta) = \frac{1}{\left(\sqrt{2\pi}\right)^n \det \mathbf{A}} \cdot \exp\left(-\frac{1}{2} \left\| \mathbf{A}^{-1} \cdot (\mathbf{x} - \langle \mathbf{x} \rangle) \right\|^2\right)$$

Given the distribution described by θ, the expectation $J(\theta)$ of the fitness becomes:

$$J(\theta) = E(f(\mathbf{x})|\theta) = \int f(\mathbf{x}) p(\mathbf{x}|\theta) d\mathbf{x}$$

The gradient of the expectation $J(\theta)$, $\nabla_\theta J(\theta)$, can be calculated by using the so-called *log-likelihood trick* according to

$$\nabla_\theta J(\theta) = \int \left(f(\mathbf{x}) \nabla \log\left(p(\mathbf{x}|\theta) \right) \right) p(\mathbf{x}|\theta) d\mathbf{x},$$

which can be approximated by Monte Carlo estimation based on the offspring individuals $\mathbf{x}_i, i \in \{1, \dots, \lambda\}$:

$$\nabla_\theta J(\theta) \approx \frac{1}{\lambda} \sum_{i=1}^{\lambda} f(\mathbf{x}_i) \nabla \log\left(p(\mathbf{x}|\theta) \right).$$

For calculating the term $\nabla \log\left(p(\mathbf{x}|\theta) \right)$, we refer to [67]. Combining this with the *Fisher information matrix* (FIM) $\mathbf{F} \in \mathbb{R}^{N \times N}$, where $N = n + n(n+1)/2$, the natural gradient G is obtained as:

$$G = \mathbf{F}^{-1} \nabla_\theta J(\theta)$$

Use of G is motivated by the fact that it is invariant with respect to linear transformations, so that the gradient converges in a correlated search space pretty much like in an isotropic one.

The NES suffer from the disadvantage of their impracticable computational complexity of $O(n^6)$, caused by the explicit calculation of the FIM and its inversion. In contrast, the xNES do not require an explicit calculation of the FIM anymore. Based on using a so-called *exponential parameterization* (see Sect. 4.1 in [26]) a transformation of θ into *natural coordinates* (see Sect. 4.2 in [26]) is applied. Using step size δ and Cholesky factor $\mathbf{B}$, an offspring $\mathbf{x}$ is then generated from the parent $\langle \mathbf{x} \rangle$ according to:

$$\mathbf{x} = \langle \mathbf{x} \rangle + \delta \mathbf{B} \mathbf{z} \text{ where } \mathbf{z} = N(\mathbf{0}, \mathbf{I}) \tag{2.18}$$

Similar to weighted recombination, the xNES uses so-called *utility values* u_i. This approach is also called *fitness shaping* in the context of an xNES. Using the rank i given by the fitness values, utility values are calculated as follows:

$$u_i = \frac{\max\left(0, \log\left(\frac{\mu}{2} + 1\right) - \log(i)\right)}{\sum_{j=1}^{\mu} \max\left(0, \log\left(\frac{\mu}{2} + 1\right) - \log(i)\right)} - \frac{1}{\lambda}$$

Using the mutation vectors z_i from Eq. 2.18, the gradients G_M for the covariance matrix and G_δ for the current search point are defined by:

$$G_M = \frac{1}{2} \sum_{i=1}^{\lambda} u_i \left(z_i z_i^T - I\right)$$

$$G_\delta = \sum_{i=1}^{\lambda} u_i z_i$$

For calculating the gradients, all λ offspring individuals are taken into account, i.e., a selection in the classical sense is not applied. Using those gradients and the learning rates η_x, η_σ and η_B, the new search point $\langle x \rangle'$, the new step sizes σ', and the new Cholesky factor B' are calculated:

$$\langle x \rangle' = \langle x \rangle + \eta_x \cdot G_\delta$$

$$\sigma' = \sigma \cdot \exp\left(\frac{\eta_\sigma}{2n} \cdot \text{tr}\left(\sum_{i=1}^{\lambda} u_i \cdot \left(z_i z_i^T - I\right)\right)\right)$$

$$B' = B \cdot \exp\left(\frac{\eta_B}{2} \cdot G_M\right)$$

Here, the exponential function of a matrix A is defined by $\exp(A) = \sum_{n=0}^{\infty} \frac{A^n}{n!}$, see [26].

The resulting pseudocode of the xNES is given in Algorithm 2.17. The default parameters of the exogenous strategy parameters are as follows:

$$\lambda = 4 + \lfloor 3 \log(n) \rfloor$$

$$\eta_x = 1$$

$$\eta_\sigma = \frac{3}{5} \cdot \frac{3 + \log(n)}{n \sqrt{n}}$$

$$\eta_B = \eta_\sigma$$

Algorithm 2.17 xNES

---

 initialize $\langle \mathbf{x} \rangle$
 $\mathbf{B} \leftarrow \mathbf{I}$
 $\sigma \leftarrow \sqrt[d]{|\det \mathbf{B}|}$
 for $i = 1 \to \lambda$ **do**
  $u_i \leftarrow \dfrac{\max\left(0, \log\left(\frac{\lambda}{2}+1\right) - \log(i)\right)}{\sum_{j=1}^{\lambda} \max\left(0, \log\left(\frac{\lambda}{2}+1\right) - \log(i)\right)} - \dfrac{1}{\lambda}$
 end for
 repeat
  for $i = 1 \to \lambda$ **do**
   $\mathbf{z}_i \leftarrow N(\mathbf{0}, \mathbf{I})$
   $\mathbf{x}_i \leftarrow \langle \mathbf{x} \rangle + \sigma \mathbf{B} \mathbf{z}_i$
  end for
  sort $\{(\mathbf{z}_i, \mathbf{x}_i)\}$ by $f(\mathbf{x}_i)$
  $\mathbf{G}_\delta \leftarrow \sum_{i=1}^{\lambda} u_i \cdot \mathbf{z}_i$
  $\mathbf{G}_M \leftarrow \sum_{i=1}^{\lambda} u_i \cdot \left(\mathbf{z}_i \mathbf{z}_i^T - \mathbf{I}\right)$
  $G_\sigma \leftarrow \operatorname{tr}(\mathbf{G}_M)/n$
  $\mathbf{G}_B \leftarrow \mathbf{G}_M - G_\sigma \cdot \mathbf{I}$
  $\langle \mathbf{x} \rangle \leftarrow \langle \mathbf{x} \rangle + \eta_x \cdot \sigma \mathbf{B} \cdot \mathbf{G}_\delta$
  $\sigma \leftarrow \sigma \cdot \exp\left(G_\sigma \cdot \frac{\eta_\sigma}{2}\right)$
  $\mathbf{B} \leftarrow \mathbf{B} \cdot \exp\left(\mathbf{G}_B \cdot \frac{\eta_B}{2}\right)$
 until termination criterion satisfied

2.2.2.11 (1+1)-Active-CMA-ES

Extending the (1+1)-Cholesky-CMA-ES with the idea of the Active-CMA-ES to take information of unsuccessful offspring into account for covariance matrix adaptation consequently leads to the development of a hybrid, the (1+1)-Active-CMA-ES [2]. Instead of using an explicit covariance matrix $\mathbf{C} = \mathbf{A}\mathbf{A}^T$, the (1+1)-Active-CMA-ES works directly with the Cholesky factor $\mathbf{A}$ and its inverse $\mathbf{A}^{-1}$. The update of $\mathbf{A}$ has been defined previously, based on Theorem 2.2.1. In order to use $\mathbf{A}^{-1}$, an extended version of this theorem is required, which we state below (without proof, see [2]):

Theorem 2.2.2. *Let* $\mathbf{C} \in \mathbb{R}^{n \times n}$ *be a symmetric, positive definite matrix with Cholesky decomposition* $\mathbf{C} = \mathbf{A}\mathbf{A}^T$, *and let* $\mathbf{C}' = \alpha \mathbf{C} + \beta \mathbf{v}\mathbf{v}^T$ *be an update transformation of* $\mathbf{C}$ *where* $\mathbf{v} \in \mathbb{R}^n \setminus \{\mathbf{0}\}$, $\alpha \in \mathbb{R}^+$ *and* $\beta \in \mathbb{R}$. *Let* $\mathbf{w} = \mathbf{A}^{-1}\mathbf{v}$ *with* $\alpha + \beta \|\mathbf{w}\|^2 > 0$ *and let* $\mathbf{C}' = \mathbf{A}'\mathbf{A}'^T$ *be the Cholesky decomposition of the updated matrix* $\mathbf{C}'$. *Then, the Cholesky factor* $\mathbf{A}'$ *and its inverse* $\mathbf{A}'^{-1}$ *are obtained as follows:* $\mathbf{A}' = \sqrt{\alpha}\mathbf{A} + \dfrac{\sqrt{\alpha}}{\|\mathbf{w}\|^2}\left(\sqrt{1 + \dfrac{\beta}{\alpha}\|\mathbf{w}\|^2} - 1\right)\mathbf{A}\mathbf{w}\mathbf{w}^T$ *and* $\mathbf{A}'^{-1} =$

$\dfrac{1}{\sqrt{\alpha}}\mathbf{A}^{-1} - \dfrac{1}{\sqrt{\alpha}\|\mathbf{w}\|^2}\left(1 - \dfrac{1}{\sqrt{1 + \beta\|\mathbf{w}\|^2/\alpha}}\right)\mathbf{w}\mathbf{w}^T\mathbf{A}^{-1}$.

The offspring $\mathbf{x}'$ is generated from its parent $\mathbf{x}$ according to:

$$\mathbf{x}' = \mathbf{x} + \sigma \mathbf{A}\mathbf{z} \text{ where } \mathbf{z} = N(\mathbf{0}, \mathbf{I})$$

As for the (1+1)-Cholesky-CMA-ES, the success rate p_s, i.e., the fraction of successful mutations, is updated by taking the learning rate c_p into account:

$$p'_s = \begin{cases} (1 - c_p)p_s + c_p & \text{if } f(\mathbf{x}') \leq f(\mathbf{x}) \\ (1 - c_p)p_s & \text{if } f(\mathbf{x}') > f(\mathbf{x}) \end{cases}$$

Based on the success rate p_s, a damping parameter $d \in \mathbb{R}^+$ and the target success rate p_t, the global step size σ is updated as follows:

$$\sigma' = \sigma \cdot \exp\left(\frac{1}{d}\frac{p_s - p_t}{1 - p_t}\right)$$

The algorithm uses $p_t = \frac{2}{11}$ which makes the update similar to the 1/5-success rule update mechanism of the (1+1)-ES.

If the offspring performs better than its parent, a positive Cholesky update is applied. In contrast to the (1+1)-Cholesky-CMA-ES, which uses the mutation step $\mathbf{z}$ for this update, the (1+1)-Active-CMA-ES relies on a search path $\mathbf{s}$, accumulating successful mutation steps with a learning rate c and updating $\mathbf{s}$ as follows:

$$\mathbf{s}' = (1 - c)\mathbf{s} + \sqrt{c(2 - c)}\mathbf{A}\mathbf{z}$$

With a constant $c_c^+ > 0$ and the vector $\mathbf{w} = \mathbf{A}^{-1}\mathbf{s}$, the positive update of matrices $\mathbf{A}$ and $\mathbf{A}^{-1}$ can now be defined according to Theorem 2.2.2:

$$\mathbf{A}' = a\mathbf{A} + b(\mathbf{A}\mathbf{w})\mathbf{w}^T \text{ and} \tag{2.19}$$

$$\mathbf{A}^{-1'} = \frac{1}{a}\mathbf{A}^{-1'} - \frac{b}{a^2 + ab\|\mathbf{w}\|^2}\mathbf{w}(\mathbf{w}^T\mathbf{A}^{-1}) \text{ where} \tag{2.20}$$

$$a = \sqrt{1 - c_c^+} \text{ and}$$

$$b = \frac{\sqrt{1 - c_c^+}}{\|\mathbf{w}\|^2}\left(\sqrt{1 + \frac{c_c^+}{1 - c_c^+}\|\mathbf{w}\|^2} - 1\right)$$

In the case of an Active-CMA-ES, the $\lambda - \mu$ worst individuals are used for the negative update of the covariance matrix, and these individuals can be called the "especially bad" individuals. In the case of the corresponding (1+1)-strategy, as introduced here, this definition is not applicable. Instead, the (1+1)-Active-CMA-ES stores past function evaluations and defines an individual to be "especially bad", if its fitness value is worse than the fitness of its k-th predecessor. For an "especially bad" offspring, a negative update according to Eqs. 2.19 and 2.20 is performed, using modified values of the coefficients a and b. In contrast to the positive update, rather than the transformed search path $\mathbf{w} = \mathbf{A}^{-1}\mathbf{s}$ the vector $\mathbf{z}$ is used for the negative update:

$$a = \sqrt{1 + c_c^-}$$

$$b = \frac{\sqrt{1 + c_c^-}}{\|\mathbf{z}\|^2} \left(\sqrt{1 - \frac{c_c^-}{1 - c_c^-} \|\mathbf{z}\|^2} - 1 \right)$$

To ensure a positive definite covariance matrix, $1 - \frac{c_c^-}{1+c_c^-}\|\mathbf{z}\|^2 > 0$ needs to hold for the constant c_c^-. Moreover, the convergence behavior of the algorithm can become unstable if the value of $1 - \frac{c_c^-}{1+c_c^-}\|\mathbf{z}\|^2$ is very close to zero. As a countermeasure, in case of $1 - \frac{c_c^-}{1+c_c^-}\|\mathbf{z}\|^2 < 1/2$, the value of c_c^- is provided with an upper bound of $1/(2\|\mathbf{z}\|^2)$.

The default settings of the exogenous parameters are:

$$d = 1 + n/2$$

$$c = 2/(n+2)$$

$$c_p = 1/12$$

$$p_t = 2/11$$

$$c_c^+ = \frac{2}{n^2 + 6}$$

$$c_c^- = \frac{2}{5(n^{8/5} + 1)}$$

The pseudocode of the (1+1)-Active-CMA-ES is given in Algorithm 2.18.

2.2.2.12 $(\mu/\mu_W, \lambda_{iid} + \lambda_m)$-ES

The $(\mu/\mu_W, \lambda_{iid} + \lambda_m)$-ES [7] is based on extending the idea of *mirrored sampling*, as described in Sect. 2.2.2.9 for a $(1 + \lambda_m^s)$-ES, for the case $\mu > 1$. The offspring population size is given by the number of samples λ_{iid} (independent, identically distributed samples from the mutation distribution) and the number of offspring, λ_m ($\lambda_m \leq \lambda_{iid}$), which are also subject to mirroring. Using *mirrored sampling* in combination with weighted recombination and cumulative step size adaptation (see Sect. 2.2.2.1) introduces a *bias* with respect to the step size, i.e., the step size is more than desirably reduced, thus potentially causing a premature stagnation effect for the algorithm. To avoid this issue, the concept of *pairwise selection* is introduced, i.e., it is made sure that recombination will not involve an offspring individual and its mirrored version at the same time, but either one or the other.

The $(\mu/\mu_W, \lambda_{iid} + \lambda_m)$-ES introduces two different versions of mirroring, namely *random mirroring* and *selective mirroring*. In the case of *random mirroring*, denoted by $(\mu/\mu_W, \lambda_{iid} + \lambda_m^{rand})$-ES, the λ_m offspring subject to mirroring are randomly selected out of the total number of offspring, λ_{iid}. In the case of *selective*

Algorithm 2.18 $(1+1)$-Active-CMA-ES

---

initialize $\mathbf{x}, \sigma, \mathbf{A} \leftarrow \mathbf{I}, \mathbf{A}^{-1} \leftarrow \mathbf{I}, \mathbf{h} \leftarrow \mathbf{0} \in \mathbb{R}^k$
$t \leftarrow 0$
repeat
    $t \leftarrow t+1$
    $\mathbf{z} \leftarrow N(\mathbf{0}, \mathbf{I})$
    $\mathbf{y} \leftarrow \mathbf{x} + \sigma \mathbf{A} \mathbf{z}$
    if $t > k$ **then**
        $h_i \leftarrow h_{i+1} \ \forall i \in \{1, \ldots, k-1\}$
        $h_k \leftarrow f(\mathbf{y})$
    else
        $h_t \leftarrow f(\mathbf{y})$
    end if
    if $f(\mathbf{y}) \leq f(\mathbf{x})$ **then**
        $\mathbf{x} \leftarrow \mathbf{y}$
        $p_s \leftarrow (1-c_p)p_s + c_p$
        $\mathbf{s} \leftarrow (1-c)\mathbf{s} + \sqrt{c(2-c)}\mathbf{A}\mathbf{z}$
        $\mathbf{w} \leftarrow \mathbf{A}^{-1}\mathbf{s}$
        $a \leftarrow \sqrt{1-c_c^+}$
        $b \leftarrow \frac{\sqrt{1-c_c^+}}{\|\mathbf{w}\|^2}\left(\sqrt{1 + \frac{c_c^+}{1-c_c^+}\|\mathbf{w}\|^2} - 1\right)$
        $\mathbf{A} \leftarrow a\mathbf{A} + b\,(\mathbf{A}\mathbf{w})\,\mathbf{w}^T$
        $\mathbf{A}^{-1} \leftarrow \frac{1}{a}\mathbf{A}^{-1} - \frac{b}{a^2+ab+\|\mathbf{w}\|^2}\mathbf{w}\,(\mathbf{w}^T\mathbf{A}^{-1})$
    else
        $p_s \leftarrow (1-c_p)p_s$
        if $h_0 < f(\mathbf{y})$ **then**
            $a \leftarrow \sqrt{1+c_c^-}$
            $b \leftarrow \frac{a}{\|\mathbf{z}\|^2}\left(\sqrt{1 - \frac{c_c^-}{1+c_c^-}\|\mathbf{z}\|^2} - 1\right)$
            $\mathbf{A} \leftarrow a\mathbf{A} + b\,(\mathbf{A}\mathbf{w})\,\mathbf{w}^T$
            $\mathbf{A}^{-1} \leftarrow \frac{1}{a}\mathbf{A}^{-1} - \frac{b}{a^2+ab+\|\mathbf{w}\|^2}\mathbf{w}\,(\mathbf{w}^T\mathbf{A}^{-1})$
        end if
    end if
    $\sigma \leftarrow \sigma \exp\left(\frac{1}{d}\frac{p_s - p_t}{1-p_t}\right)$
until termination criterion satisfied

---

mirroring, denoted by $(\mu/\mu_W, \lambda_{iid} + \lambda_m^{sel})$-ES, the λ_{iid} offspring are first sorted by fitness and the λ_m worst individuals undergo mirroring. This approach is motivated by considering that, in a convex objective function topology, mirroring the best offspring cannot yield any further improvement, such that it will be advantageous to mirror the worst individuals. Moreover, since bad offspring in the case of a (μ_W, λ)-ES are often generated by applying too-large mutation steps, *selective mirroring* itself will also favor large mutation steps [7]. To counteract this undesired bias, the *resample length* approach changes the length of the mirrored mutation step by additionally using a second, newly sampled mutation vector $\mathbf{z}'$. The mirrored version $\mathbf{x}_m$ of the offspring $\mathbf{x} = \langle \mathbf{x} \rangle + \sigma \mathbf{z}$ is then created according to $\mathbf{x}_m = \langle \mathbf{x} \rangle - \sigma \frac{\|\mathbf{z}'\|}{\|\mathbf{z}\|}\mathbf{z}$.

Like for the $(1 \overset{+}{,} \lambda_m^s)$-ES, theoretical results for the convergence velocity on the sphere model have been derived, see [7]. In particular, it has been shown that, for

---

Algorithm 2.19 $(\mu/\mu_W, \lambda_{iid} + \lambda_m)$-ES

---

  initialize $\langle \mathbf{x} \rangle, \sigma$
  $r \leftarrow 0$
  repeat
    $i \leftarrow 0$
    while $i < \lambda_{iid}$ **do**
      $r \leftarrow r + 1$
      $i \leftarrow i + 1$
      $\mathbf{x}_i \leftarrow \langle \mathbf{x} \rangle + \sigma N(\mathbf{0}, \mathbf{I})$
    end while
    if *selective mirroring* **then**
      $\mathbf{x}_1, \ldots, \mathbf{x}_{\lambda_{iid}} = \text{argsort}\left(f(\mathbf{x}_1), \ldots, f(\mathbf{x}_{\lambda_{iid}})\right)$
    end if
    while $i < \lambda_{iid} + \lambda_m$ **do**
      $i \leftarrow i + 1$
      if *resample length* **then**
        $r \leftarrow r + 1$
        $\mathbf{x}_i \leftarrow \langle \mathbf{x} \rangle - \frac{\sigma \|N(\mathbf{0},\mathbf{I})\|}{\|\mathbf{x}_{i-\lambda_m} - \langle \mathbf{x} \rangle\|}\left(\mathbf{x}_{i-\lambda_m} - \langle \mathbf{x} \rangle\right)$
      else
        $\mathbf{x}_i \leftarrow \langle \mathbf{x} \rangle - \left(\mathbf{x}_{i-\lambda_m} - \langle \mathbf{x} \rangle\right)$
      end if
    end while
    $\mathbf{x}_1, \ldots, \mathbf{x}_{\lambda_{iid}} = \text{argsort}(f(\mathbf{x}_1), \ldots, f(\mathbf{x}_{\lambda_{iid}-\lambda_m}),$
        $\min\{f(\mathbf{x}_{\lambda_{iid}-\lambda_m+1}), f(\mathbf{x}_{\lambda_{iid}+1})\}, \ldots,$
        $\min\{f(\mathbf{x}_{\lambda_{iid}}), f(\mathbf{x}_{\lambda_{iid}+\lambda_m})\}\})$
    $\sigma \leftarrow \text{updateStepSize}(\sigma, \mathbf{x}_1, \ldots, \mathbf{x}_{\lambda_{iid}}, \langle \mathbf{x} \rangle)$
    $\langle \mathbf{x} \rangle \leftarrow \langle \mathbf{x} \rangle + \sum_{i=1}^{\mu} w_i (\mathbf{x}_i - \langle \mathbf{x} \rangle)$
  until termination criterion satisfied

---

the sphere model, maximum convergence velocity is achieved for a setting of $r = \lambda_m/\lambda_{iid} \approx 0.1886$, which can serve as a guideline for the fraction of offspring individuals which should be mirrored.

The pseudocode as given in Algorithm 2.19 is based on using a method *updateStepSize*[26] to update the step size σ, and weights $w_i \; \forall i \in \{1, \ldots, \mu\}$, such that $\sum_{i=1}^{\mu} w_i = 1$.

2.2.2.13 SPO-CMA-ES

The SPO-CMA-ES [70] is essentially a restart-version of the (μ_W, λ)-CMA-ES. It is based on using *sequential parameter optimization* (SPO) [11] to optimize the exogenous parameters of an evolution strategy. SPO uses methods of *design of experiments* (DoE) and *design and analysis of computer experiments* (DACE).[27]

---

[26]The aforementioned techniques self-adaptation (see Sect. 2.2.1.2) or cumulative step size adaptation (see Sect. 2.2.2.1) are suitable.

[27]See [70] for literature references on these topics as well as the Kriging modeling method.

Concerning the exogenous parameters subject to sequential parameter optimization, the number of offspring individuals[28] $\lambda \in \{\lambda_{def}, \dots, 1{,}000\}$, the initial step size $\sigma_{init} \in [1, 5]$ and the so-called *selection pressure* $\lambda/\mu \in [1.5, 2.5]$ are identified.

The pseudocode of the SPO-CMA-ES is provided in Algorithm 2.20, and the approach is explained in the following by discussing the various methods used in the algorithm. To begin with, using *latin hypercube sampling* (LHS) [68] an initial design of experiments for the exogenous parameters is created. In the next step (runDesign), independent runs of the (μ_W, λ)-CMA-ES are executed, using the parameter sets of the DoE plan. The results, i.e., the best evaluated individual with its fitness value, of each run is collected in the set Y. This initial phase of the algorithm is called the *exploration phase*.

The next phase, called the *exploitation phase*, is repeated until the predefined budget of function evaluations is reached. Using a function aggregateRuns, a performance measure y is calculated for every run configuration in Y. Based on these performance measure values as outputs and the corresponding parameter sets according to the experimental plan, a Kriging model[29] $\mathcal{M}$ is trained in the method fitModel. This Kriging model $\mathcal{M}$ is then used by the method modelOptimization to identify a new design point, e.g., by running an optimization on the Kriging model and using the resulting point. The new design point d is then added to the experimental plan D, and the loop is executed again. Default settings are not given for the size of the initial experimental plan, N_{init}, nor for the split of the number of function evaluations between the two phases of the algorithm [70]. Rather, the user of the algorithm can fix them, depending on the optimization task at hand. In the case of noisy objective functions, the method runDesign can execute more than the one run, in order to use, e.g., the averages as an estimation of the true fitness value.

2.3   Further Aspects of ES

So far, we have described the ES algorithms as single-criterion optimizers with $\mathbb{R}^n$ as search domain and without handling of constraints. The next three sections give summarized overviews and literature references for further aspects of ES, namely constraint handling, binary and integer search spaces, and multiobjective optimization.

---

[28]For λ_{def} the standard setting of a (μ_W, λ)-CMA-ES with $\lambda_{def} = 4 + \lfloor 3 \log n \rfloor$ is used.

[29]In principal, any modeling technique can be used to establish the relationship between the exogenous parameters and the performance measure.

Algorithm 2.20 SPO-CMA-ES

Input: box constraints $\mathbf{l}, \mathbf{u} \in \mathbb{R}^n$ and size N_{init} of the initial design
Output: final model $\mathcal{M}$ and best design point d^*

$\quad i \leftarrow 0, D \leftarrow \emptyset$
$\quad d_i \leftarrow \text{LHS}(\mathbf{l}, \mathbf{u}, N_{init})$
$\quad Y \leftarrow \text{runDesign}(d_i)$
$\quad D \leftarrow D \cup d_i$
$\quad$**while** function evaluation budget not exhausted **do**
$\quad\quad i \leftarrow i + 1$
$\quad\quad y \leftarrow \text{aggregateRuns}(Y)$
$\quad\quad \mathcal{M} \leftarrow \text{fitModel}(D, y)$
$\quad\quad d_i \leftarrow \text{modelOptimization}(\mathcal{M})$
$\quad\quad Y \leftarrow Y \cup \text{runDesign}(d_i)$
$\quad\quad D \leftarrow D \cup d_i$
$\quad$**end while**
$\quad d^* \leftarrow d_k$ with the best $y_k \in \{y_0, \ldots, y_i\}$

2.3.1 Constraint Handling

In Sect. 2.1.1 we defined the optimization problem used throughout this book with equality and inequality constraints as in Eq. 2.2. There are many techniques for handling constraints ranging from simple penalty methods to more complex ones like hybrid methods involving Lagrangian multipliers. Coello gives an overview [18] of constraint-handling techniques to be used with Evolutionary Algorithms but some of these methods may be applied to ES as well. A review by Kramer [42] specializes in constraint-handling methods dedicated to ES and presents the four techniques *penalty methods*, a *multiobjective bioinspired approach*, *coordinate alignment techniques*, and *metamodeling of constraints*.

2.3.2 Beyond Real-Valued Search Spaces

There are many optimization problems where the search domain is not constrained to the real domain. Especially decision problems[30] use categorical search spaces, in most cases binary search spaces, i.e., $\mathbf{x} \in \{0, 1\}^n$, as the simplest categorical search space. Another search space of practical use is the integer search space representable as a subset of $\mathbb{Z}$. Originally, Genetic Algorithms (see [27] or [25] for a comprehensive introduction) were designed to handle binary search spaces, but there are approaches to incorporate those search spaces into ES. In Sect. 2.1.3 we named three guidelines to choose a distribution to be used for mutation. Rudolph [56] introduces a mutation operator for integer search spaces using the difference

---

[30]For example the NP-hard Traveling Salesman Problem.

of two geometrical distributions. Each discrete variable of a categorical search space is assigned a probability whether to mutate or not. The new value of the discrete variable is drawn uniformly from all possible values. The MI-ES (*mixed-integer* evolution strategies) [43] solve optimization problems which are mixed in their search domain, i.e. the search domain is a composition of real, integer and categorical search spaces. They use the aforementioned mutation approaches together with self-adaptation for the endogenous parameters. An overview of other approaches for handling mixed search spaces is given by Li [43].

2.3.3 Multiobjective Optimization

In single-objective optimization fitness values can be ordered to decide whether one solution is better than another. In multiobjective optimization, where fitness values are represented as vectors, such a strict ordering does not exist anymore. Solutions are partially ordered and based on the partial order solutions can be either *dominated* or *non-dominated* by other solutions. Hence there is not a single optimum to be found but a set of solutions which is called the *Pareto set* or *Pareto front*. For a detailed description of these concepts see [20]. Algorithms for multiobjective optimization have to measure how well a Pareto front is approximated. The most common measures for this task are the *crowding distance* and the *hypervolume contribution*. The former is used for example by NSGA-II [21] the latter by SMS-EMOA [12].

Chapter 3
Taxonomy of Evolution Strategies

In order to provide an integrated overview of the various developments in modern evolution strategies, this chapter provides a possible taxonomy and classification of the algorithms. Section 3.1 starts by providing the different development strands of evolution strategies. In Sect. 3.2, characteristics of modern evolution strategies are identified which can be used for defining the corresponding taxonomy. Finally, based on the properties of modern evolution strategies, practical recommendations for their usage, depending on the particular application area, are provided in Sect. 3.3.

3.1 Development Strands of Modern Evolution Strategies

3.1.1 Overview

Before discussing the development strands of modern evolution strategies in detail, this section briefly identifies these development strands, starting with a general historical overview as shown in Table 3.1. Even though this list is still not complete, it contains a few more algorithms than we described in Chap. 2. In particular, three evolution strategies using so-called meta-modeling approaches (i.e., approaches to approximating the fitness function topology by means of data-driven modeling algorithms), namely the lmm-CMA-ES, nlmm-CMA-ES, and p-sep-lmm-CMA-ES are mentioned in the list due to the fact that these algorithms are directly derived from the CMA-ES variants. However, the area of meta-modelling itself is quite extensive, and therefore we have decided to omit this topic here, i.e., these algorithms are neither described in Chap. 2, nor are they taken into account in the empirical comparison presented in Chap. 4. However, their fundamental principles are briefly described in Sect. 3.1.4.

Towards our goal of providing a taxonomy, Fig. 3.1 provides a summary of the development strands of modern evolution strategies 1996–2011. The arrows in

T. Bäck et al., *Contemporary Evolution Strategies*, Natural Computing Series, DOI 10.1007/978-3-642-40137-4_3, © Springer-Verlag Berlin Heidelberg 2013

Table 3.1 Historical development of key evolution strategies

Year of publication	Name	Key reference(s)
1964	(1+1)-ES	[59]
1973	(1+1)-ES 1/5-success rule	[52]
1977	$(1,\lambda)$-ES	[61]
1975–1981	$(\mu/\rho \overset{+}{,} \lambda)$-ES	[60–62]
1981	(μ,λ)-MSC-ES	[62]
1993	DR1	[47]
1994	DR2	[48]
1995	DR3	[33]
1996	(μ,λ)-CMA-ES	[31]
2004	LS-CMA-ES	[6]
2005	LR-CMA-ES	[4]
2005	IPOP-CMA-ES	[5]
2006	(1+1)-Cholesky-CMA-ES	[38]
2006	Active-CMA-ES	[40]
2006	lmm-CMA-ES	[41]
2008	(μ,λ)-CMSA-ES	[13]
2008	NES	[71]
2008	sep-CMA-ES	[54]
2009	eNES	[66]
2010	$(1 \overset{+}{,} \lambda_m^s)$-ES	[16]
2010	xNES	[26]
2010	(1+1)-Active-CMA-ES	[2]
2010	nlmm-CMA-ES	[14]
2011	$(\mu_W, \lambda_{iid} + \lambda_m)$-ES	[7]
2011	SPO-CMA-ES	[70]
2011	p-sep-lmm-CMA-ES	[15]

the diagram indicate advancements of algorithms. As the figure reveals, almost all algorithms are based on the (μ_W, λ)-CMA-ES, with the exceptions of the family of *natural evolution strategies*, i.e., NES, eNES and xNES, and the two algorithms $(\mu_W, \lambda_{iid} + \lambda_m)$-ES and $(1 \overset{+}{,} \lambda_m^s)$-ES. The former, *natural evolution strategies*, provide an independent approach to covariance matrix adaptation, while the latter, $(\mu_W, \lambda_{iid} + \lambda_m)$-ES and $(1 \overset{+}{,} \lambda_m^s)$-ES, are more general methods for offspring creation and selection, which can also be applied to other variants of evolution strategies than the (μ_W, λ)-CMA-ES.

Concerning a potential classification, the modern evolution strategies discussed in this book can be separated into three main groups described in the following sections, namely:

1. Restart heuristics (see Sect. 3.1.2).
2. Methods for adaptation of mutation parameters (see Sect. 3.1.3).
3. Methods for avoiding function evaluations (see Sect. 3.1.4).

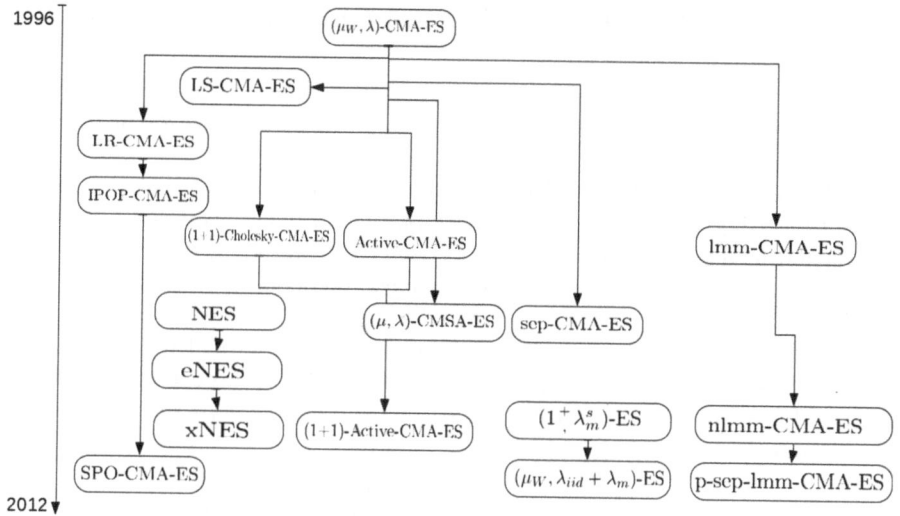

Fig. 3.1 Development strands of modern evolution strategies

3.1.2   Restart Heuristics

In the case of multimodal objective functions, evolution strategies like any other optimization algorithm run the risk of getting trapped in a local optimum. In such a situation, restart heuristics can provide a decent approach for identifying a number of different local optima through restarts and then choosing the best of the local optima found. In general, restart heuristics are applied to an evolution strategy as an outer loop, executing a run of the evolution strategy until stagnation is observed, and then starting a new run, typically based on modified initial parameters for the evolution strategy run. The execution of runs is repeated until a termination criterion, e.g., reaching the maximum number of runs, is satisfied.

Among the modern evolution strategies described here, the LR-CMA-ES, IPOP-CMA-ES, and SPO-CMA-ES are instances of restart heuristics.

The LR-CMA-ES (see Sect. 2.2.2.3) uses criteria for discovering stagnation, and applies them to the (μ_W, λ)-CMA-ES. Except for the initial search point, none of the parameters of the algorithm are modified between the different runs.

The IPOP-CMA-ES (see Sect. 2.2.2.4) increases the population size for each new run, and, for the first time, changes exogenous strategy parameters between the runs.

The SPO-CMA-ES as described in Sect. 2.2.2.13 represents a broad generalization of the IPOP-CMA-ES, using information from stagnated runs by applying changes to exogenous strategy parameters which are predicted by a meta-model to yield improved success of the evolution strategy runs. The underlying idea of *sequential parameter optimization* can be applied in principle to any kind of evolution strategy, provided that ranges of the exogenous strategy parameters subject to optimization are defined.

3.1.3 Methods for Adapting Mutation Parameters

All of the modern evolution strategies discussed are based on using the multivariate normal distribution for mutation, and the mutation parameters are defined by step sizes and covariances.

In the case of the (μ_W, λ)-CMA-ES, the corresponding covariance matrix is adapted by using the evolution path (*rank-one-update*) and a matrix $\mathbf{Z}$ which is generated from the μ best mutation vectors (*rank-μ-update*). The LS-CMA-ES uses two different modes for covariance matrix adaptation, namely, on the one hand the adaptation algorithm of the (μ_W, λ)-CMA-ES, and on the other hand learning the Hessian $\mathbf{H}$ from past samples[1] by *least squares* estimation. Provided the estimation $\hat{\mathbf{H}}$ of the Hessian reaches a certain quality level, $\mathbf{C} = \frac{1}{2}\hat{\mathbf{H}}^{-1}$ is used as the covariance matrix. This approach suffers from the fact that solving the *least squares* problem takes computational effort $O(n^6)$.

Another alternative for covariance matrix adaptation is based on incorporating negative updates when generating matrix $\mathbf{Z}$, as introduced by the Active-CMA-ES. This approach increases convergence velocity by taking not only the μ best mutation vectors into account for calculating $\mathbf{Z}$, but also in addition the $\lambda - \mu$ worst ones, with negative sign.

Cholesky updates are an extension of the general covariance matrix adaptation with the aim of reducing computational complexity from $O(n^3)$ to $O(n^2)$. This was first implemented within the $(1 + 1)$-Cholesky-CMA-ES, by dropping the explicit usage of the covariance matrix $\mathbf{C}$ and instead adapting the Cholesky factor $\mathbf{A}$, defined by $\mathbf{C} = \mathbf{AA}^T$.

The $(1 + 1)$-Active-CMA-ES combines these two approaches, i.e., taking also the $\lambda - \mu$ worst mutation vectors into account and using the Cholesky update, thus benefiting from both the convergence velocity speedup as well as the quadratic rather than cubic computational effort for mutation.

The xNES and its predecessors NES and eNES use a different approach for adapting endogenous mutation parameters. They are based on using a so-called *natural gradient*, which defines a direction towards better fitness in the space of mutation parameters. The relationship between the adaptation mechanisms of a (μ_W, λ)-CMA-ES and an xNES are described in Sect. 4.4 of [26].

The CMSA-ES reintroduces self-adaptation for the global step size again, and it adapts the covariance matrix similarly to a *rank-one update* by using a vector which is adapted by (equally weighted) intermediary recombination of the best mutation steps. An evolution path is not used by this strategy.

Finally, the sep-CMA-ES introduces a radical simplification of covariance matrix adaptation by restricting the covariance matrix to be a diagonal matrix. This reduces computational effort to $O(n)$ at the expense of losing the ability to generate correlated mutations; only anisotropic mutations can be generated.

---

[1] Samples are tuples of the form $(\mathbf{x}, f(\mathbf{x}))$.

3.1.4 Methods for Avoiding Function Evaluations

A method developed specifically for avoiding function evaluations is the *sequential selection* approach as used in the $(1 + \lambda_m^s)$-ES. For a plus-strategy, offspring individuals are sequentially evaluated upon their creation, and as soon as an improvement over the parent individual has been found, offspring creation and evaluation is stopped.

Another approach relies on *meta-modeling* techniques, using algorithms from data-driven modeling (see, e.g., [45] or [37]) to compute a predictive approximation model of the objective function, based on already evaluated individuals. The model prediction can then be used rather than a real function evaluation, provided that the predictive quality of the model is sufficient. Estimating the accuracy of a prediction is the key issue in meta-modeling, and typically cross-validation approaches are used to estimate the prediction accuracy of such models. A common example is *leave-one-out* cross-validation as used in [43], where of k available data points $k - 1$ are used for model training and the remaining one for evaluating the prediction. Repeating this process systematically for all k points provides a fairly good measure of model quality. However, cross-validation is computationally quite expensive, so the *meta-modeling* evolution strategy algorithms[2] lmm-CMA-ES [41], nlmm-CMA-ES [14], and p-sep-lmm-CMA-ES [15] use a different approach. They are all based on the (μ_W, λ)-CMA-ES and, due to the weighted recombination, do not require exact function values, but just a prediction of the ranking of offspring. To achieve such a ranking, the lmm-CMA-ES introduces a so-called *approximate ranking procedure*, which however does not prove to achieve the desired reduction in the number of function evaluations. The nlmm-CMA-ES develops this concept further by using a less constrained assessment of model quality.

3.2 Characteristics of Modern Evolution Strategies

3.2.1 Computational Effort

We use the term computational effort to denote the computational complexity, depending on the dimensionality n, of the evolution strategy within a single generation. Typically, its dominating component is given by the numerical adaptation of the endogenous strategy parameters. For most of the modern evolution strategies, an eigendecomposition of the covariance matrix $\mathbf{C}$ needs to be computed, which causes computational complexity of $O(n^3)$.

---

[2]lmm, nlmm, and p-sep-lmm are abbreviations for *local meta model*, *new local meta model*, and *partially separable local meta model*.

Table 3.2 Computational
complexity of modern
evolution strategies,
for a single generation

Algorithm	Computational complexity
(μ_W, λ)-CMA-ES	$O(n^3)$
LS-CMA-ES	$O(n^6)$
$(1 + 1)$-Cholesky-CMA-ES	$O(n^2)$
Active-CMA-ES	$O(n^3)$
(μ, λ)-CMSA-ES	$O(n^3)$
sep-CMA-ES	$O(n)$
xNES	$O(n^3)$
$(1 + 1)$-Active-CMA-ES	$O(n^2)$

Exceptions with a smaller complexity include the $(1 + 1)$-Cholesky-CMA-ES, the $(1 + 1)$-Active-CMA-ES and the sep-CMA-ES. Concerning the LS-CMA-ES, a clearly higher and typically prohibitive complexity of $O(n^6)$ is required; see Chap. 2 for an explanation of the reasons causing this. A brief summary of the corresponding adaptation algorithms for endogenous strategy parameters is provided in Sect. 3.1.3, and Table 3.2 gives an overview of the computational complexity of the key variants of modern evolution strategies—excluding the restart heuristics, since they depend on the underlying evolution strategy used for restarts.

3.2.2 Convergence Behavior

In this section, we provide a short summary of relevant results of empirical investigations of modern evolution strategies concerning their convergence behavior, following the results discussed in the original literature. In most cases, the corresponding algorithm was compared to the reference algorithm, the (μ_W, λ)-CMA-ES. In contrast to the empirical analysis discussed in Chap. 4, the results in the original literature are based on a very large number of function evaluations.

According to the experiments discussed in [6], the LS-CMA-ES requires about a factor of three to four times fewer function evaluations than the (μ_W, λ)-CMA-ES in the case of elliptic functions, to reach the same level of convergence. Empirical investigations reported in [40] show that for all objective functions tested, except for f_1 (see Table 4.1), the Active-CMA-ES performs better than the (μ_W, λ)-CMA-ES. This is an indication of the general advantage of using negative updates for covariance matrix adaptation for convergence behavior of an evolution strategy. Moreover, negative updates seem to be of particular advantage in cases where the eigenvalue spectrum of the Hessian of the objective function is dominated by a single large eigenvalue [40].

Another experimental investigation of the (μ, λ)-CMSA-ES, published in [13], demonstrates that the (μ, λ)-CMSA-ES outperforms the (μ_W, λ)-CMA-ES for growing population sizes on a number of test functions. These tests, however, have been executed using $\lambda = 4n^2$, which is far larger than the default setting $\lambda = 4 + \lfloor 3 \ln n \rfloor$ of the (μ_W, λ)-CMA-ES.

For the sep-CMA-ES, experiments clarified an improved convergence behavior on separable functions [54], although the number of runs executed (11 for $n < 100$ and 2 for $n \gg 100$) seems to be too small general conclusions about its convergence behavior to be drawn.

An empirical investigation of the xNES illustrates clear improvements compared to its predecessor, the eNES, but compared to the (μ_W, λ)-CMA-ES it can only improve convergence behavior in the case of a few functions for $n = 2$ (a setting of n which we would consider irrelevant for any practical applications) [26].

In the case of restart heuristics, the comparison between LR-CMA-ES and IPOP-CMA-ES illustrates a general advantage of increasing the population size, with a few exceptions (namely, test functions 13, 21, and 23 in [65]).

For the SPO-CMA-ES, results indicate this algorithm performs better than the IPOP-CMA-ES as the number of restarts increases over time [70].

3.3 Recommendations for Practical Use

3.3.1 Global Optimization

In the case of global optimization tasks, i.e., trying to find the best one of a large set of local optima, a restart heuristic should be used. Among those restart heuristics discussed in this book, the SPO-CMA-ES provides the most general approach. It uses information from previous runs which stagnated in local optima in order to adapt the exogenous strategy parameters. This approach is more flexible than the steady increase of population size used in the IPOP-CMA-ES. Concerning the evolution strategy used within the SPO-CMA-ES, according to the results presented in Sect. 3.2.2, the Active-CMA-ES should be used. Alternatively, for larger population sizes, the (λ, μ)-CMSA-ES seems to be the best choice.

3.3.2 High-Dimensional Search Spaces

In the case of high-dimensional search spaces, i.e., $n \gg 100$, an evolution strategy requiring a small computational effort should be used. The computational complexity of the adaptation of endogenous strategy parameters is discussed in detail in Sect. 3.2.1.

Based on those results, the sep-CMA-ES with linear effort clearly provides the lowest effort approach, however, at the expense of losing the ability to generate correlated mutations, clearly causing a severe loss of convergence velocity. Since correlated mutations are often or even typically required, the

(1+1)-Active-CMA-ES is a natural choice, due to its quadratic computational effort, which is still much better than the typical cubic complexity.

In general, however, the balance between computational effort and convergence properties depends on the computational effort of the function evaluations. If a function evaluation requires much more time than the adaptation of endogenous strategy parameters, evolution strategies with cubic complexity can be used again.[3]

---

[3]For example, on a state-of-the-art computer (Intel Core i7-2600 3.4 GHz), Octave requires a few seconds for the eigendecomposition of a $1,000 \times 1,000$ matrix, and more than 3 min in the case of a $5,000 \times 5,000$ matrix.

Chapter 4
Empirical Analysis

One goal of this book is to empirically answer the question of how efficient ES are in a setting of few function evaluations with a focus on modern ES from Sect. 2.2.2. This chapter addresses the experiments conducted and is organized as follows. Section 4.1 introduces two measures to evaluate the efficiency of ES, the *fixed cost error* (FCE) and the *expected run time* (ERT). Section 4.2 describes how the experiments were conducted technically and how they are examined. The results are presented and discussed in Sect. 4.3.

4.1 Measuring Efficiency

An ES is considered efficient in this book if it approaches the optimum f^* (see Eq. 2.4) quickly, i.e., by using as few function evaluations as possible. In order to compare different ES, a measure of efficiency for the convergence properties of an ES is needed. Figure 4.1 shows a sample convergence plot for five optimization runs[1] of an ES. The x-axis of the convergence plot represents the number of function evaluations. The y-axis shows the base ten logarithm of the difference between the currently best function value and the optimum f^*. This difference will be called Δf^* in the following. For a plus-strategy the graph is monotonically decreasing. To achieve monotonicity for a comma-strategy as well, one uses the best evaluated individual found so far for the calculation of Δf^* instead of the best individual of the current generation.

Appendix D.3 in [34] describes two opposing approaches for deriving an efficiency measure from these convergence plots. On the one hand there is the *fixed-cost view*, on the other hand there is the *fixed-target view*. The *fixed-cost view*

---

[1]Actually, these runs were five independent runs of the (μ, λ)-MSC-ES on the 10-dimensional *sphere function* (f_1 in BBOB).

T. Bäck et al., *Contemporary Evolution Strategies*, Natural Computing Series, DOI 10.1007/978-3-642-40137-4_4, © Springer-Verlag Berlin Heidelberg 2013

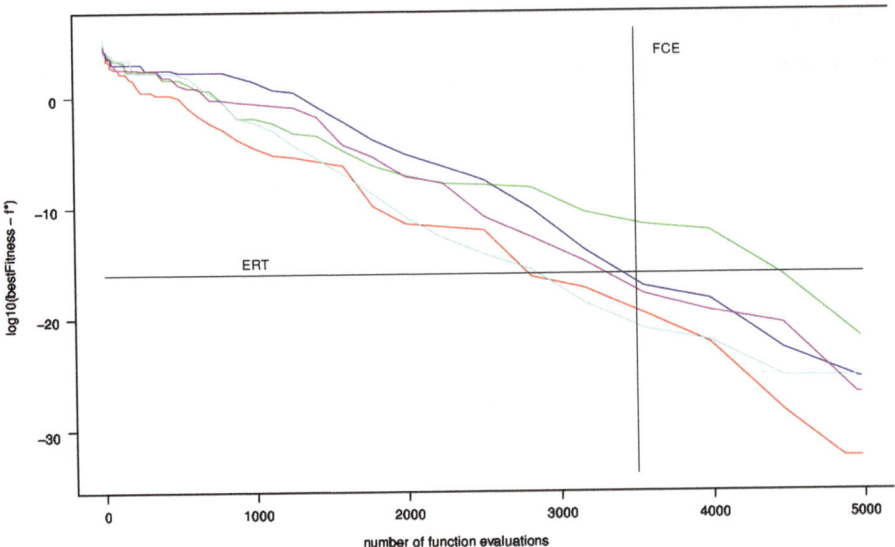

Fig. 4.1 Example of a convergence plot

operates with a fixed number of function evaluations and yields the *fixed-cost error* (FCE) covered in detail in Sect. 4.1.1. The other approach leads to the *expected runtime* measure (ERT) which is used in the BBOB benchmarking framework and described in Sect. 4.1.2.

4.1.1   The FCE Measure

FCE measures Δf^* given a fixed number of function evaluations. Considering the convergence plot in Fig. 4.1 this approach is implemented graphically by drawing a vertical line. The FCE values are represented by the intersections between the convergence graphs and the vertical line. FCE is of relevance for industrial applications demanding a maximal run-time hence a fixed number of function evaluations. In [34] the relevance of FCE is acknowledged but the use of FCE is rejected because it does not allow for a quantitative interpretation. The lack of interpretation stems from the fact that the ratio between two Δf^* from two algorithms cannot quantify how much better one algorithm is than another. Nevertheless, a qualitative interpretation is possible. On the basis of FCE one can answer the question of which algorithm yields a smaller FCE and as a result is better. Since optimization runs with ES are influenced by random effects, both during initialization and running period, the FCE of an algorithm has to be measured based on many independent optimization runs. Then, the FCE of different algorithms can be analyzed with statistical techniques to find significant differences in quality. This analysis is described in Sect. 4.2.3.

4.1.2 The ERT Measure

BBOB uses ERT as the measure for benchmarking algorithms. It was introduced in [49] as *expected number of function evaluations per success* and further analyzed under the name *success probability* (SP) [4]. ERT is the expected number of function evaluations needed to reach a given Δf^*. Graphically, ERT consists of an intersection between a convergence graph as shown in Fig. 4.1 and a horizontal line representing a fixed Δf^*. With this approach there might be optimization runs which do not reach the given Δf^* within a finite amount of function evaluations. These runs are considered unsuccessful and are rated with the run-time r_{us}. A successful run is rated with the number of function evaluations to reach Δf^*, i.e., the run-time r_s. The ratio of successful runs to all runs yields the value p_s. Let R_s and R_{us} be the mean value of the r_s and r_{us} from different runs, then for $p_s > 0$ ERT is defined as:

$$\text{ERT} = R_s + \frac{1 - p_s}{p_s} R_{us}$$

If there are unsuccessful runs, i.e., $p_s < 1$, then ERT strongly depends on the termination criteria of the algorithm and the value of r_{us}. Considering optimization runs with very few function evaluation easily leads to $p_s = 0$ when using common values for Δf^*. So, to use ERT in this scenario, appropriate values for Δf^* have to be found first.

4.2 Experiments

4.2.1 Selection of Algorithms

Not all modern ES algorithms covered in Sect. 2.2.2 were subject to an empirical analysis. Since the focus of this book is on optimization runs with very few function evaluations, the restart heuristics were omitted. They can be better analyzed by long-running optimizations. Interesting results from such runs conducted by the authors of the algorithms are summarized in Sect. 3.2.2. In addition, five algorithms developed before 1996, described in Sect. 2.2.1, are included in the experiments. The complete list of algorithms which were used in the experiments is shown in Table 4.3 in Sect. 4.2.2.2.

4.2.2 Technical Aspects

4.2.2.1 Framework

The experiments are performed using the framework BBOB 10.2 [34]. It provides standardized test functions and interfaces to the programming languages C, C++, Java and Matlab/Octave. Having an implementation of an algorithm in one of these languages allows us to conduct experiments with minimal organizational effort on a set of test functions F, a set of function instances FI and a set of dimensions D for the input space. The set FI controls the number of independent runs for a test function. Let $R = F \times D \times FI$, then $|R|$ runs are conducted in total. A run, i.e., an element of R, yields a table indexed by the number of function evaluations used, containing information regarding the optimization run. This information is the difference between the current noise free fitness and the optimum and the difference between the best measured[2] noise free fitness and the optimum. For small dimensionality the input values **x** yielding the current fitness are displayed as well. BBOB provides Python scripts for post-processing these tables.

Runs are conducted on all 24 noise free test functions. A detailed description of the test functions can be found on the BBOB web page.[3] The global optima of all test functions are located inside the hyperbox $[-5, 5]^n$. The test functions can be classified by different features. A test function can be uni- or multimodal, i.e., having only one or multiple (local) optima. Multimodal functions allow the global optimization capabilities of an algorithm to be benchmarked. Furthermore, a test function can be symmetric, i.e., invariant under rotations of the coordinate system. The condition of a function can be interpreted as a reciprocal measure of its symmetry und depends on the condition of an optimal covariance matrix (see Sect. 2.2.2.2). A more vivid description is that a function with a high condition has a fitness landscape with very steep valleys. Table 4.1 provides a summary of all 24 test functions with their commonly used names and some of their features.

Considering their features, test functions can be classified. The discrimination into separable and non-separable and unimodal and multimodal functions are of special interest. Table 4.2 shows this distribution of test functions across these four classes. Unimodal test functions have a unique optimum which make them suitable for testing convergence properties of an algorithm without interferences stemming from stagnation in local optima. Multimodal test functions are especially useful for testing restart heuristics or algorithms designed to escape a local optimum. Since real fitness functions are usually multimodal, multimodal functions comply better with real-world optimization scenarios than unimodal functions.

Separable functions allow the optimization run to be split into n independent one-dimensional optimizations. In contrast to this, non-separable functions cannot be

---

[2]For a plus-strategy these two values are the same.

[3]http://coco.gforge.inria.fr/doku.php?id=bbob-2010-downloads

Table 4.1 BBOB test functions

Symbol	Name	Features
f_1	*Sphere*	Unimodal, highly symmetric
f_2	*Ellipsoidal*	Unimodal, separable, condition $>10^6$
f_3	*Rastrigin*	Multimodal (about 10^n local optima)
f_4	*Büche-Rastrigin*	Multimodal (about 10^n local optima), less symmetric than f_3
f_5	*Linear slope*	Unimodal, the optimum is located on the edge of the search space
f_6	*Attractive sector*	Unimodal, highly asymmetric
f_7	*Step ellipsoidal*	Unimodal with many plateaus
f_8	*Rosenbrock*	Multimodal
f_9	*Rotated Rosenbrock*	Multimodal
f_{10}	*Ellipsoidal*	Unimodal, non-separable version of f_2
f_{11}	*Discus*	Unimodal, condition $>10^6$
f_{12}	*Bent cigar*	Unimodal, condition $>10^6$
f_{13}	*Sharp ridge*	Unimodal, non-differentiable near the optimum
f_{14}	*Different powers*	Unimodal, highly sensitive area in the vicinity of the optimum
f_{15}	*Rastrigin*	Multimodal, non-separable version of f_3
f_{16}	*Weierstrass*	Multimodal without unique global optimum
f_{17}	*Schaffer's F7*	Highly multimodal
f_{18}	*Ill-conditioned Schaffer's F7*	Highly multimodal with greater condition than f_{17}
f_{19}	*Composite Griewank-Rosenbrock*	Highly multimodal
f_{20}	*Schwefel function*	Multimodal
f_{21}	*Gallagher's Gaussian 101-me peaks*	Multimodal with randomly distributed local optima
f_{22}	*Gallagher's Gasussian 21-hi peaks*	Multimodal with randomly distributed local optima
f_{23}	*Katsuura*	Highly multimodal with very steep valleys
f_{24}	*Lunacek bi-Rastrigin*	Highly multimodal

Table 4.2 Classification of test functions

	Separable	Non-separable
Unimodal	$\{1, 2, 5\}$	$\{6, 7, 10, 11, 12, 13, 14\}$
Multimodal	$\{3, 4\}$	$\{8, 9, 15, 16, 17, 18, 19, 20, 21, 22, 23, 24\}$

Table 4.3 Summary of ES implementations

ES	Implementation
$(1 + 1)$-ES	Own implementation
(μ, λ)-MSC-ES	Own implementation
DR1	Own implementation
DR2	Own implementation
DR3	Own implementation
(μ_W, λ)-CMA-ES	cmaes.m version 3.55beta by N. Hansen
LS-CMA-ES	Own implementation
$(1 + 1)$-Cholesky-CMA-ES	Own implementation
Active-CMA-ES	cmaes.m version 3.55beta by N. Hansen
(μ, λ)-CMSA-ES	Own implementation
sep-CMA-ES	Own implementation
$(1 + 1)$-Active-CMA-ES	Own implementation
$(1, 4_m^s)$-CMA-ES	cmaes.m version 3.41beta modified by A. Auger
xNES	xnes.m by Y. Sun

optimized this way and for them it is advantageous to apply an ES using correlated mutations. In general, non-separable multimodal functions are far more difficult to solve and hence serve better as an idealization of real-world problems.

According to [34], 15 runs are sufficient to observe significant differences when comparing 2 algorithms. For the analysis based on the FCE measure a *best-of-n* approach (described in Sect. 4.2.3) is used. In order to observe significant differences with this approach as well, the number of runs per test function, defined by the *function instances* in BBOB, is increased to 100. BBOB recommends a maximum number of function evaluations of $10^6 \cdot n$. Since the focus is on optimization tasks allowing only very few function evaluations, a drastically decreased maximum number of function evaluations of $500 \cdot n$ is chosen. The experiments are conducted with dimensions $n \in \{2, 5, 10, 20, 40, 100\}$. For the dimensions $n = 40$ and $n = 100$ the maximum number of function evaluations is reduced to 10^4. The initial search point is drawn uniformly from the hyperbox $[-5, 5]^n$.

4.2.2.2 Software for ES Algorithms

The BBOB framework is used with its interface to the Matlab/Octave programming language. If there are publicly available implementations[4] by the authors of the ES algorithms, they are used. For most of the ES an original implementation was created. Table 4.3 provides an overview of the implementation used.

---

[4]N. Hansen and A. Auger's CMA-ES is available at https://www.lri.fr/~hansen/cmaes_inmatlab. html; Y. Sun's xNES is available at http://www.idsia.ch/~tom/code/xnes.m.

All original implementations[5] represent the pseudocode of the algorithms, as listed in Chap. 2, for Octave [23]. Furthermore, these implementations are capable of constraining the search space to a hyperbox (see Eq. 2.5). For this purpose a transformation as described in [43] is applied individually to the coordinates of a search point. The transformed value x' of $x \in \mathbb{R}$ subject to the lower bound l and the upper bound u is calculated as follows:

$$x' = l + (u - l)\frac{2}{\pi}\sin^{-1}(|\sin\left(\frac{\pi(x - l)}{2(u - l)}\right)|)$$

Simply speaking, the transformation performs a reflection at the bounds. An optimization run is terminated if either the maximum number of function evaluations is reached or the fitness falls below a given target value. These two values can be configured by parameters in all the original implementations. The exogenous parameters of the different ES algorithms are configured with their default settings as described in Sect. 2.2.

4.2.3 Analysis

In the following, the procedure for evaluating the empirical test results is outlined.

4.2.3.1 Calculating FCE from Empirical Results

The basis for the calculation of FCE is the tables described in Sect. 4.2.2.1, which are called BBOB data in the following. The BBOB data contains tuples $(\#fe, \Delta f^*)$, which consist of $\#fe$, the number of function evaluations performed, and Δf^*, the so-far best[6] difference from the optimum f^*. There is not necessarily a tuple for every $\#fe \in \{1, \ldots, \#fe_{max}\}$ in the BBOB data. Let $I \subset \{1, \ldots, \#fe_{max}\}$ be the subset of existing $\#fe$ values in the BBOB data with C_t as target costs, then FCE is calculated as follows:

$$\text{FCE}(C_t) = \Delta f^* \text{ from the tuple } (\#fe, \Delta f^*) \text{ with } \#fe = \sup\{e | e \in I \wedge e \leq C_t\}$$

Simply speaking, the FCE for a specific C_t is based on the closest C_t available in the BBOB data, which is smaller than or equal to the desired C_t. In this way, the performance of an algorithm might be underestimated but is not overestimated.

---

[5]The Octave source code is available for non-commercial use at the web site of divis intelligent solutions GmbH (http://www.divis-gmbh.com/es-software.html), see Sect. 1.4.
[6]This allows for comparing comma and plus strategies.

4.2.3.2   Calculating Rankings

Conducting $m = 100$ runs for each test function f and each dimension n yields a set $E(f, n, C_t)$ containing m FCE(C_t) values. For each algorithm a the sets $E(f, n, C_t)_a$ can be analyzed pairwise with non-parametric statistical tests [36] to find significant differences in their FCE. We use unpaired Welch Student's t-tests [69] to decide whether one algorithm is better than another.[7] The difference between the mean of $E(f, n, C_t)_1$ and the mean of $E(f, n, C_t)_2$ is considered significant for a *p-value* < 0.05 and the algorithm with the better mean FCE is declared the winner and gets a point. Doing so pairwise for all algorithms, the algorithms are ranked according to the number of points obtained.

In [24] two relevant optimization scenarios are described. In the first one the user has the opportunity to choose the best run out of several runs. For this purpose an algorithm with a good *peak performance*, i.e., an algorithm which performs very well sometimes but its general performance is highly variant, is appropriate. In the second scenario only one optimization run is done. This requires an algorithm to have a good performance without much variation. This kind of performance is called the *average performance* of an algorithm. To reflect these two scenarios in our analysis, we will use a *best-of-k* approach. Instead of using all m runs to create the set $E(f, n, C_t)$ only the best out of k runs can be used. This reduces the cardinality of $E(f, n, C_t)$ to $\lfloor \frac{m}{k} \rfloor$. The analysis regarding the average performance of an algorithm is done with $k = 1$. For the peak performance we have to choose an appropriate k. The resulting set $E(f, n, C_t)$ must not be too small in order to apply statistical testing for significant differences. We choose a *best-of-k* approach with $k = 5$ to rank the algorithms regarding their peak performance.

4.2.3.3   Selection of Test Functions

Until now the sets E were dependent on one test function. In order to calculate a rank aggregation for a set of test functions, the points won by an algorithm for each test function within the set are accumulated before determining the aggregated ranking. Aggregated rankings are calculated for the classes of test functions as assigned in Table 4.2.

4.2.3.4   Choice of Target Costs C_t

Following the motivation of this work small values for target costs C_t are chosen. C_t should be dependent on the dimension n to facilitate the interpretation of results for different dimensions. BBOB recommends $10^6 \cdot n$ for long runs thus

---

[7] We used the free statistics software R [50] for this purpose.

establishing a linear dependency. We choose to analyze results for $C_t = \alpha \cdot n$ with $\alpha \in \{25, 50, 100\}$ instead, i.e., our focus is on much smaller values for C_t.

4.3   Results

4.3.1   Ranks by FCE

The following figures show rankings aggregated for the four function classes as described in Sect. 4.2.2.1. Each ranking is displayed for all dimensions. Instead of using the rank, the number of significant wins over other algorithms divided by the number of test functions per class is shown on the y-axis. This kind of normalization allows the plots for different function classes to be compared. With 14 algorithms tested an algorithm can achieve at most 13 significant wins. This representation also has the advantage of showing how clearly an algorithm wins or loses against others. The aggregated ranking over all 24 test functions is given in Table 4.4.

4.3.2   Discussion of Results

Based on the results we are able to answer two questions regarding optimization scenarios with very few function evaluations. The first one is: Are there significant differences in the convergence properties of Evolutions Strategies with few function evaluations? In general this question can be answered positively. Even $25 \cdot n$ function evaluations are sufficient to observe significant differences. As can be seen in Figs. 4.2–4.7 there are hardly any significant differences in algorithm performance for non-separable, multimodal test functions with dimension $n = 2$. An explanation for this behaviour is given by the fact that the variance of the Euclidian distance between the initial search point and the global optimum in the search space decreases with the dimensionality.[8] That means for $n = 2$ the variance of the differences is relatively high and the initialization of the search point impacts the results too much for us be able to see more significant differences in the convergence behaviour of the algorithms tested. According to the ranking aggregated over all 24 test functions as shown in Table 4.4, the Active-CMA-ES is clearly the best evolution strategy for optimization scenarios with few function evaluations, followed by the (μ_W, λ)-CMA-ES in second place. This result holds

---

[8]The Euclidian distance of two points uniformly drawn from a hyper box in $\mathbb{R}^n$ is distributed according to the normal distribution $N(\sqrt{n}, 1/\sqrt{2})$ (see e.g. [53]). Hence, with increasing n the variance decreases w.r.t. the mean.

Table 4.4 Aggregated rankings over all 24 test functions for $C_t = 100 \cdot n$. Columns p show ranks for the *peak performance* (*best-of-5*) and columns a represent ranks for the *average performance* (*best-of-1*)

ES	n = 2		n = 5		n = 10		n = 20		n = 40		n = 100	
	p	a	p	a	p	a	p	a	p	a	p	a
(1 + 1)-ES	8	9	9	9	7	8	6	7	8	9	9	7
(μ, λ)-MSC-ES	10	7	13	11	13	13	13	13	11	13	11	11
DR1	5	6	6	4	6	5	9	6	7	5	10	9
DR2	6	10	5	6	4	4	4	4	3	3	3	3
DR3	12	14	12	13	12	12	11	12	12	12	12	12
(μ_W, λ)-CMA-ES	3	4	2	2	2	3	2	2	2	2	2	2
LS-CMA-ES	11	12	11	11	10	11	10	10	10	10	5	5
(1 + 1)-Cholesky-CMA-ES	7	10	7	10	9	9	7	7	9	8	7	6
Active-CMA-ES	1	2	1	1	1	1	1	1	1	1	1	1
(μ, λ)-CMSA-ES	13	13	7	8	8	6	5	5	4	3	4	4
sep-CMA-ES	14	5	14	14	14	14	14	14	14	14	14	14
(1 + 1)-Active-CMA-ES	2	8	4	7	5	7	7	9	5	7	8	7
$(1, 4_m^s)$-CMA-ES	4	3	2	3	3	2	3	3	6	6	6	10
xNES	9	1	10	5	11	9	12	11	13	11	13	13

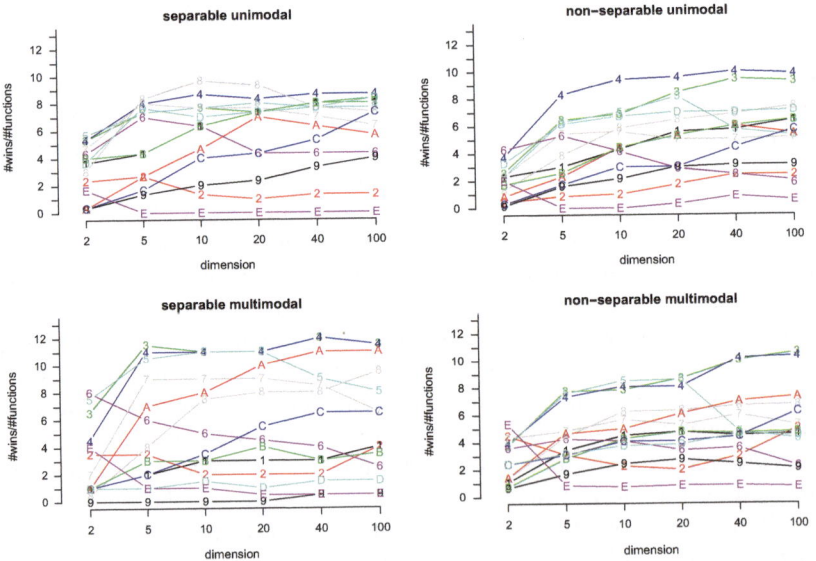

Fig. 4.2 Ranking for $C_t = 100 \cdot n$ with *best-of-1* approach. 1: (1 + 1)-ES, 2: (μ, λ)-MSC-ES, 3: (μ_W, λ)-CMA-ES, 4: Active-CMA-ES, 5: $(1, 4_m^s)$-CMA-ES, 6: xNES, 7: DR1, 8: DR2, 9: DR3, A: (μ, λ)-CMSA-ES, B: (1 + 1)-Cholesky-CMA-ES, C: LS-CMA-ES, D: (1 + 1)-Active-CMA-ES, E: sep-CMA-ES

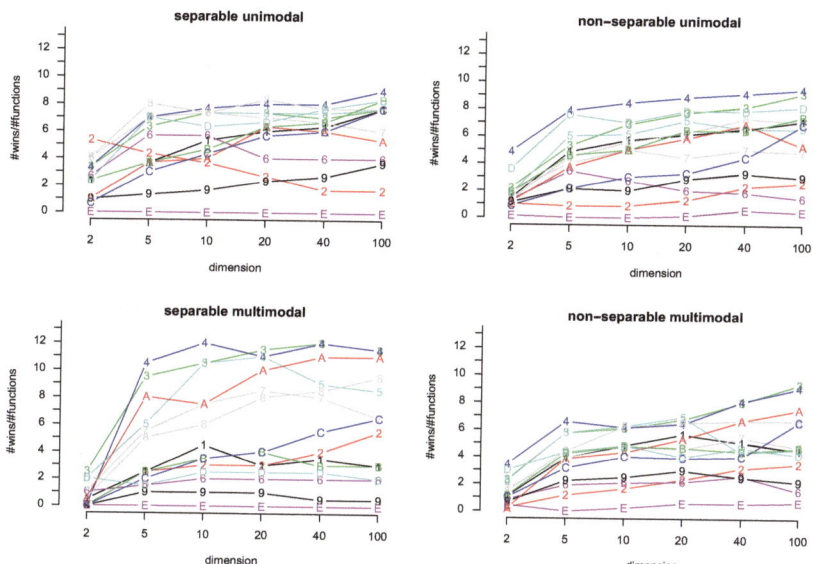

Fig. 4.3 Rankings for $C_t = 100 \cdot n$ with *best-of-5* approach. 1: $(1 + 1)$-ES, 2: (μ, λ)-MSC-ES, 3: (μ_W, λ)-CMA-ES, 4: Active-CMA-ES, 5: $(1, 4_m^s)$-CMA-ES, 6: xNES, 7: DR1, 8: DR2, 9: DR3, A: (μ, λ)-CMSA-ES, B: $(1 + 1)$-Cholesky-CMA-ES, C: LS-CMA-ES, D: $(1 + 1)$-Active-CMA-ES, E: sep-CMA-ES

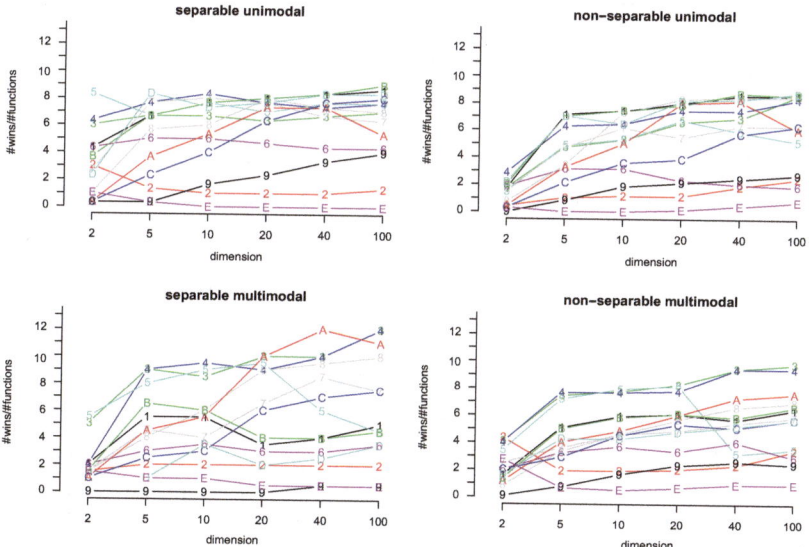

Fig. 4.4 Rankings for $C_t = 50 \cdot n$ with *best-of-1* approach. 1: $(1 + 1)$-ES, 2: (μ, λ)-MSC-ES, 3: (μ_W, λ)-CMA-ES, 4: Active-CMA-ES, 5: $(1, 4_m^s)$-CMA-ES, 6: xNES, 7: DR1, 8: DR2, 9: DR3, A: (μ, λ)-CMSA-ES, B: $(1 + 1)$-Cholesky-CMA-ES, C: LS-CMA-ES, D: $(1 + 1)$-Active-CMA-ES, E: sep-CMA-ES

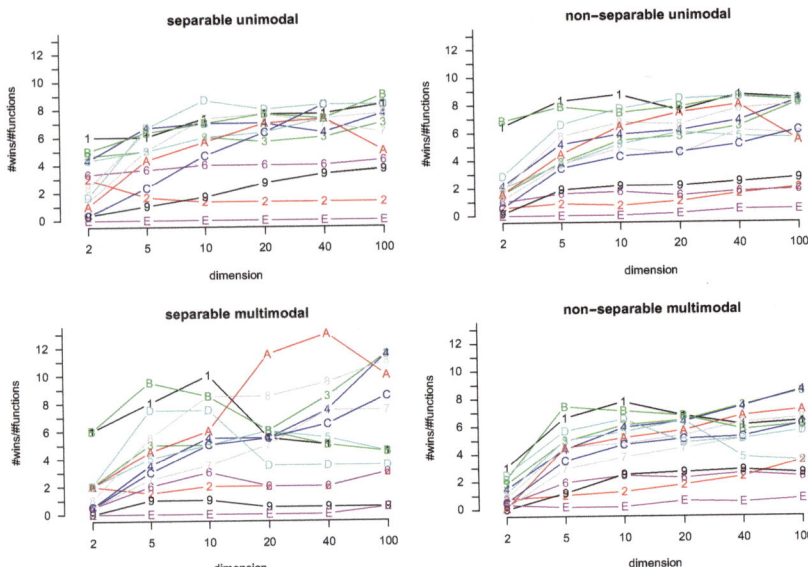

Fig. 4.5 Rankings for $C_t = 50 \cdot n$ with *best-of-5* approach. 1: $(1 + 1)$-ES, 2: (μ, λ)-MSC-ES, 3: (μ_W, λ)-CMA-ES, 4: Active-CMA-ES, 5: $(1, 4_m^s)$-CMA-ES, 6: xNES, 7: DR1, 8: DR2, 9: DR3, A: (μ, λ)-CMSA-ES, B: $(1 + 1)$-Cholesky-CMA-ES, C: LS-CMA-ES, D: $(1 + 1)$-Active-CMA-ES, E: sep-CMA-ES

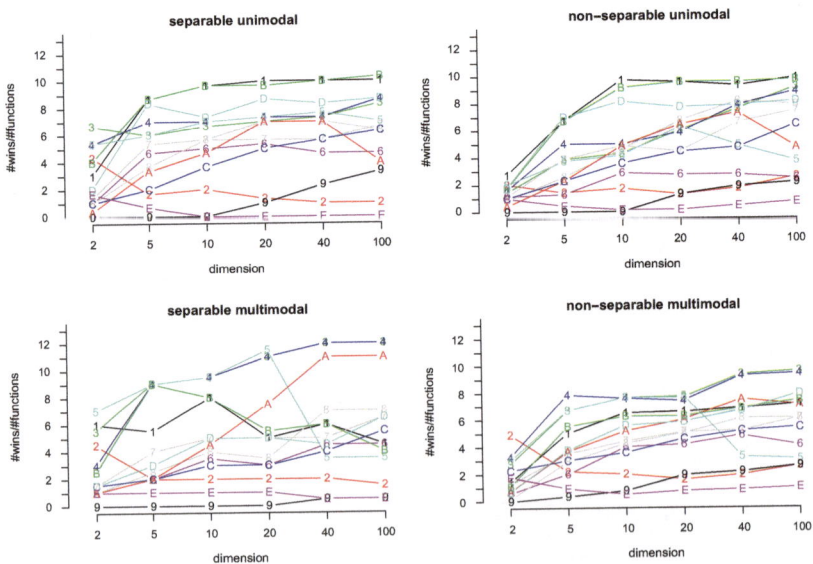

Fig. 4.6 Rankings for $C_t = 25 \cdot n$ with *best-of-1* approach. 1: $(1 + 1)$-ES, 2: (μ, λ)-MSC-ES, 3: (μ_W, λ)-CMA-ES, 4: Active-CMA-ES, 5: $(1, 4_m^s)$-CMA-ES, 6: xNES, 7: DR1, 8: DR2, 9: DR3, A: (μ, λ)-CMSA-ES, B: $(1 + 1)$-Cholesky-CMA-ES, C: LS-CMA-ES, D: $(1 + 1)$-Active-CMA-ES, E: sep-CMA-ES

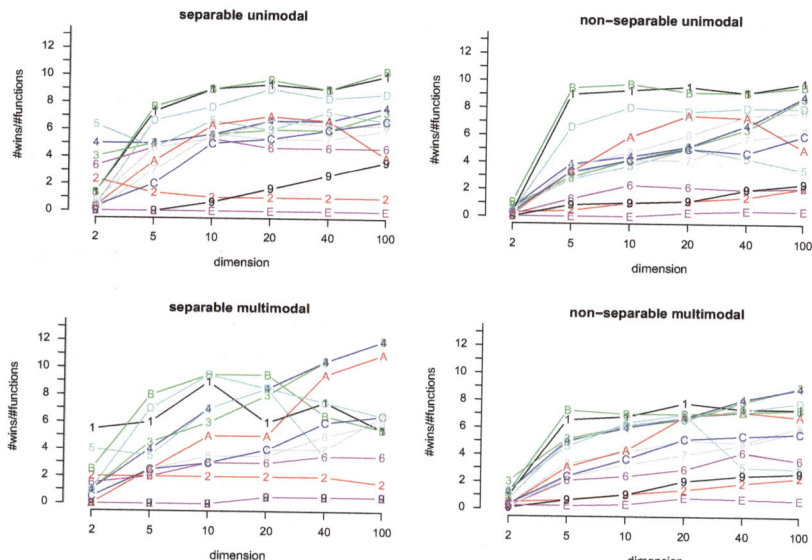

Fig. 4.7 Rankings for $C_t = 25 \cdot n$ with *best-of-5* approach. 1: $(1 + 1)$-ES, 2: (μ, λ)-MSC-ES, 3: (μ_W, λ)-CMA-ES, 4: Active-CMA-ES, 5: $(1, 4_m^s)$-CMA-ES, 6: xNES, 7: DR1, 8: DR2, 9: DR3, A: (μ, λ)-CMSA-ES, B: $(1 + 1)$-Cholesky-CMA-ES, C: LS-CMA-ES, D: $(1 + 1)$-Active-CMA-ES, E: sep-CMA-ES

regardless of whether we analyze the *peak* or *average performance*. The sep-CMA-ES clearly ranked last in these experiments.

The second question, whether there are Evolution Strategies which are better given many function evaluations but are beaten given few function evaluations, can also be answered positively in some cases. For target costs $C_t = 100 \cdot n$ the Active-CMA-ES or the (μ_W, λ)-CMA-ES usually rank best. Decreasing the target costs to $25 \cdot n$ or $50 \cdot n$ results in several $(1 + 1)$-strategies[9] being found with good rankings, especially for unimodal functions. With the *peak performance* approach the $(1 + 1)$-Cholesky-CMA-ES and the $(1 + 1)$-ES rank first, sometimes even for multimodal functions. The CMA-ES variants catch up with more function evaluations, which can be explained by the time needed to adapt the covariance matrix successfully.

Despite using only anisotropic mutations with local step sizes the DR2 algorithm performs quite well. It often ranks directly behind the successful CMA-ES variants. Thus, it offers a better alternative to the sep-CMA-ES when the runtime of the algorithm cannot be neglected w.r.t. the time for a function evaluation, which might be the case for very high dimensional search spaces.

---

[9]In detail these are the $(1+1)$-ES, the $(1+1)$-Cholesky-CMA-ES and the $(1+1)$-Active-CMA-ES.

4.4 Further Analysis for $n = 100$

As the last section illustrates, several $(1 + 1)$-ES algorithms outperform CMA-ES variants considered state of the art when it comes to very few function evaluations. In industrial optimization scenarios, where function evaluations are extremely time consuming, we are interested in quick progress rather than finding the exact global optimum, or even converging to a local optimum.

A more thorough analysis for search space dimension $n = 100$ reflecting these scenarios was also conducted. The experiments described in the last section used the performance measure FCE based on the distance to the global optimum Δf^* to quantify progress of an algorithm. In order to reflect the scenario of quick progress we chose to measure the progress made w.r.t. the initial search point instead of using Δf^* directly. So, the Δf^*_{init} of the function evaluation of the inital search point is used to normalize the Δf^* of later iterations yielding monotonically decreasing progress values.[10] Based on these values we can state by which order of magnitude an algorithm decreases the initial fitness value for a given test function after a given number of function evaluations. In order to decrease the influence of the initial search point the number of runs is increased from 100 used in the previous section to 1,000 for each of the 14 algorithms and each of the 24 test functions. As an example, Fig. 4.8 shows the resulting convergence plot for test function f_1.

As in the analysis in Sect. 4.3.1 we used the non-parametric Student's t-test to find significant differences between the algorithms tested. According to these significant differences we are able to rank the algorithms for the four test function classes shown in Table 4.2.

The results of this additional test are summarized in Tables 4.5–4.8 for the four different classes of objective functions. In addition, the corresponding convergence plots for all objective functions are provided in Figs. 4.8–4.31. The following observations can be made when analyzing the results:

- As clarified by the rankings, the $(1 + 1)$-Active-CMA-ES most often ranks first, regardless of the function class (with the exception of separable multimodal functions and large values of C_t, for which DR2 is the best algorithm). In general, the $(1 + 1)$-algorithms, even including the simple $(1 + 1)$-ES, perform quite well. It seems that adapting endogenous search parameters in the beginning more frequently with less information is better than less frequently with more information as is the case in population-based strategies.
- On non-separable, multimodal test functions, the $(1 + 1)$-Active-CMA-ES is the clear winner, followed by the $(1 + 1)$-ES. Similar performance can be observed for the other function classes.

---

[10]Monotonicity for comma-strategies can be guaranteed by using the so-far best Δf^* instead of the Δf^* of the current iteration.

Table 4.5 Aggregated rankings over the separable unimodal test functions for target costs $C_t = \{100, 200, \ldots, 1{,}000\}$

ES	100	200	300	400	500	600	700	800	900	1,000
$(1 + 1)$-ES	1	2	2	2	2	2	2	2	2	2
(μ, λ)-MSC-ES	9	11	11	11	11	12	12	12	12	12
(μ_W, λ)-CMA-ES	10	7	7	4	4	5	5	6	6	6
Active-CMA-ES	10	9	9	6	6	6	7	8	8	8
$(1, 4_m^s)$-CMA-ES	8	7	7	9	9	9	9	9	9	6
XNES	12	10	10	7	6	4	4	4	4	9
DR1	4	5	4	8	8	8	7	5	4	4
DR2	6	4	5	5	6	6	6	6	6	6
DR3	14	14	14	14	14	14	14	14	14	14
CMSA-ES	13	13	13	12	11	11	11	11	11	11
$(1 + 1)$-Cholesky-CMA-ES	2	2	3	2	2	2	2	2	2	2
LS-CMA-ES	5	6	6	10	10	10	10	10	10	10
$(1 + 1)$-Active-CMA-ES	2	1	1	1	1	1	1	1	1	1
sep-CMA-ES	7	11	12	13	13	13	13	13	13	13

Table 4.6 Aggregated rankings over the non-separable unimodal test functions for target costs $C_t = \{100, 200, \ldots, 1{,}000\}$

ES	100	200	300	400	500	600	700	800	900	1,000
$(1 + 1)$-ES	1	2	2	2	2	2	2	2	2	2
(μ, λ)-MSC-ES	8	10	10	10	11	12	12	12	12	12
(μ_W, λ)-CMA-ES	10	9	8	9	8	8	8	8	8	10
Active-CMA-ES	10	8	8	8	8	9	8	9	9	9
$(1, 4_m^s)$-CMA-ES	12	12	13	12	12	11	10	10	10	8
XNES	10	7	7	6	6	6	6	6	6	7
DR1	4	5	5	5	5	5	5	5	5	5
DR2	6	4	4	4	4	4	4	4	4	4
DR3	14	14	14	14	14	14	14	14	14	14
CMSA-ES	13	13	12	11	10	10	11	11	11	11
$(1 + 1)$-Cholesky-CMA-ES	3	2	2	2	2	2	2	2	2	2
LS-CMA-ES	5	6	6	7	7	7	7	7	7	6
$(1 + 1)$-Active-CMA-ES	2	1	1	1	1	1	1	1	1	1
sep-CMA-ES	7	11	11	13	13	13	13	13	13	13

- The convergence plots for the different functions indicate that, for the more complicated functions (e.g., f_{21}, f_{22}), progress in the beginning is very slow and accelerates later on. In contrast to this, on easier unimodal functions such as f_1 the algorithms generally converge much faster (up to three orders of magnitude improvement) after 1,000 function evaluations, and the progress rate is already high during the first 100 evaluations.

Table 4.7 Aggregated rankings over the separable multimodal test functions for target costs $C_t = \{100, 200, \ldots, 1,000\}$

ES	100	200	300	400	500	600	700	800	900	1,000
$(1 + 1)$-ES	2	3	3	3	3	3	3	5	5	6
(μ, λ)-MSC-ES	8	7	8	8	8	9	9	9	10	10
(μ_W, λ)-CMA-ES	12	12	12	12	12	12	12	12	13	13
Active-CMA-ES	12	12	12	12	12	12	12	12	13	13
$(1, 4_m^s)$-CMA-ES	11	12	11	11	12	10	10	9	9	9
XNES	9	7	7	7	7	5	5	2	2	2
DR1	6	5	5	5	5	5	5	5	4	3
DR2	4	4	4	4	3	3	2	1	1	1
DR3	14	14	14	14	14	14	14	14	13	13
CMSA-ES	10	10	10	9	9	8	8	8	8	8
$(1 + 1)$-Cholesky-CMA-ES	1	1	2	2	2	3	3	4	5	5
LS-CMA-ES	4	5	5	5	6	7	7	7	7	7
$(1 + 1)$-Active-CMA-ES	2	1	1	1	1	1	1	2	3	4
sep-CMA-ES	7	9	9	10	10	11	11	11	11	11

Table 4.8 Aggregated rankings over the non-separable multimodal test functions for target costs $C_t = \{100, 200, \ldots, 1,000\}$

ES	100	200	300	400	500	600	700	800	900	1,000
$(1 + 1)$-ES	2	2	3	3	4	3	3	4	4	3
(μ, λ)-MSC-ES	7	7	7	9	11	11	11	11	11	11
(μ_W, λ)-CMA-ES	10	8	8	9	9	9	9	10	10	9
Active-CMA-ES	9	8	8	9	9	10	10	9	9	10
$(1, 4_m^s)$-CMA-ES	14	14	14	14	14	14	13	13	13	12
XNES	12	11	10	7	7	7	8	8	8	8
DR1	5	5	5	5	5	5	5	3	3	3
DR2	4	3	3	2	2	2	2	2	2	2
DR3	13	13	13	13	13	13	14	14	14	14
CMSA-ES	11	12	12	11	8	8	7	7	7	7
$(1 + 1)$-Cholesky-CMA-ES	2	2	3	4	3	4	3	5	5	5
LS-CMA-ES	6	6	6	6	6	6	6	6	6	6
$(1 + 1)$-Active-CMA-ES	1	1	1	1	1	1	1	1	1	1
sep-CMA-ES	8	10	11	12	12	12	12	12	12	13

In conclusion, the $(1 + 1)$-Active-CMA-ES is a good recommendation for a small function evaluation budget (i.e., up to $10 \cdot n$) and high-dimensional problems in general. Especially for non-separable, multimodal test functions, it consistently shows the best performance, and for the unimodal functions it fails to win in

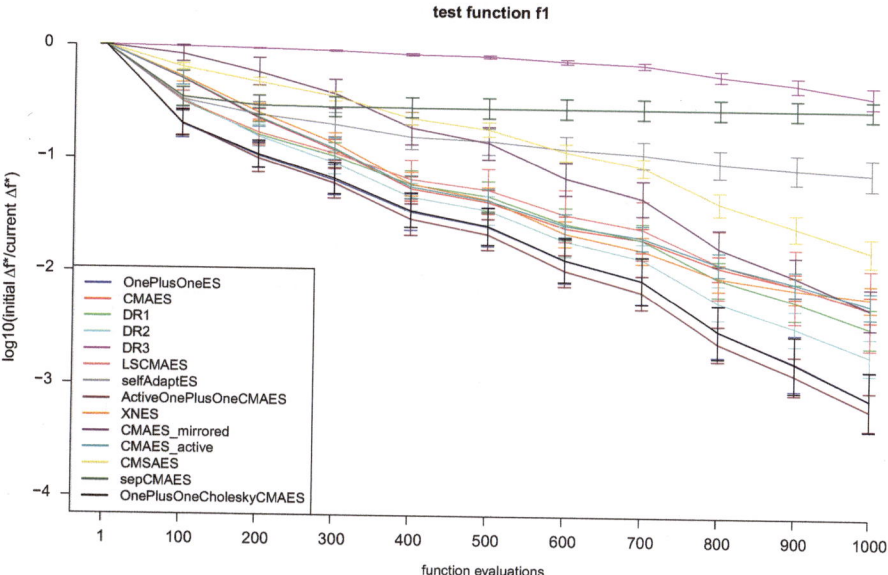

Fig. 4.8 Convergence plot for test function f_1 (sphere function) showing the order of magnitude of fitness value normalized w.r.t. the fitness of the inital search point for the number of function evaluations $\{100, 200, \ldots, 1,000\}$. Error bars reflect the 20 % respectively 80 % quantiles of the 1,000 conducted runs

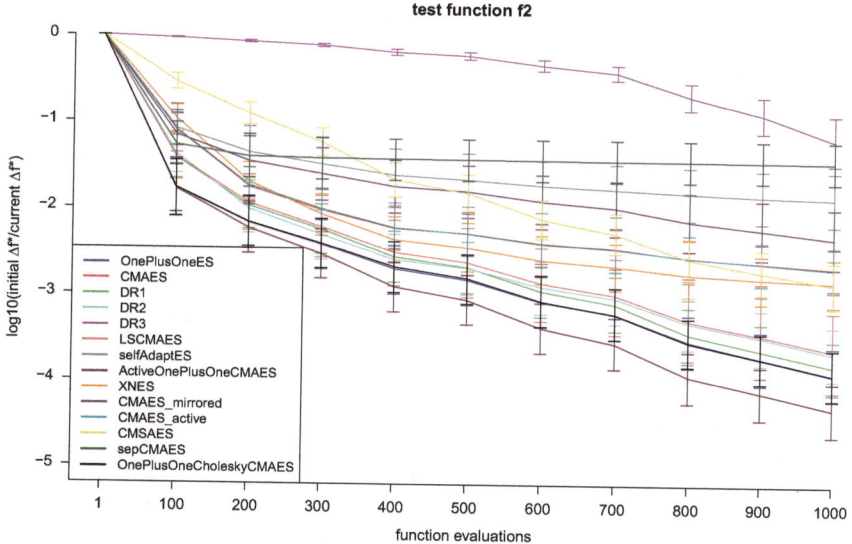

Fig. 4.9 Convergence plot for test function f_2 showing the order of magnitude of fitness value normalized w.r.t. the fitness of the inital search point for the number of function evaluations $\{100, 200, \ldots, 1,000\}$. Error bars reflect the 20 % respectively 80 % quantiles of the 1,000 conducted runs and the *solid line* represents their mean

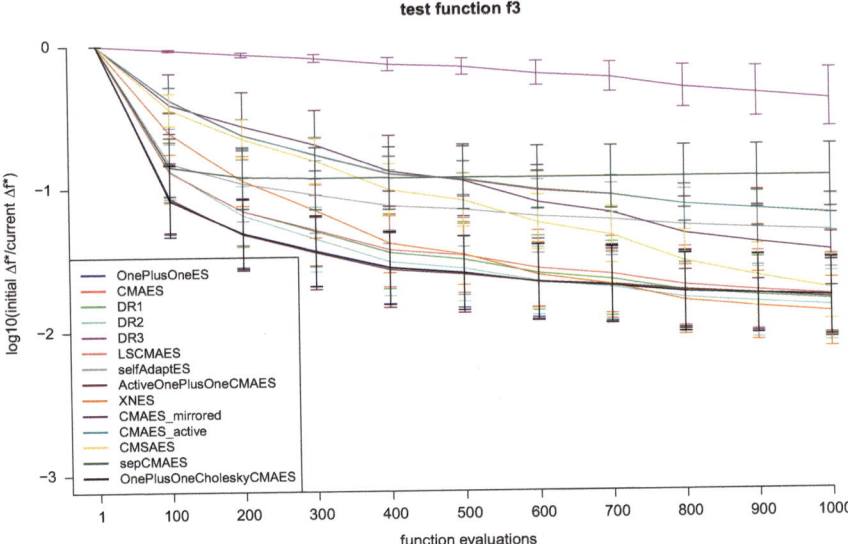

Fig. 4.10 Convergence plot for test function f_3 showing the order of magnitude of fitness value normalized w.r.t. the fitness of the initial search point for the number of function evaluations $\{100, 200, \ldots, 1,000\}$. Error bars reflect the 20 % respectively 80 % quantiles of the 1,000 conducted runs and the *solid line* represents their mean

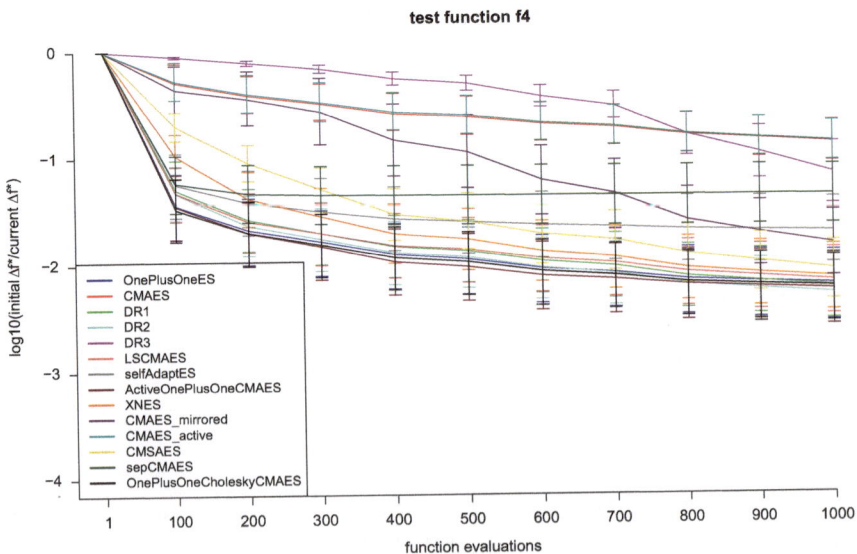

Fig. 4.11 Convergence plot for test function f_4 showing the order of magnitude of fitness value normalized w.r.t. the fitness of the initial search point for the number of function evaluations $\{100, 200, \ldots, 1,000\}$. Error bars reflect the 20 % respectively 80 % quantiles of the 1,000 conducted runs and the *solid line* represents their mean

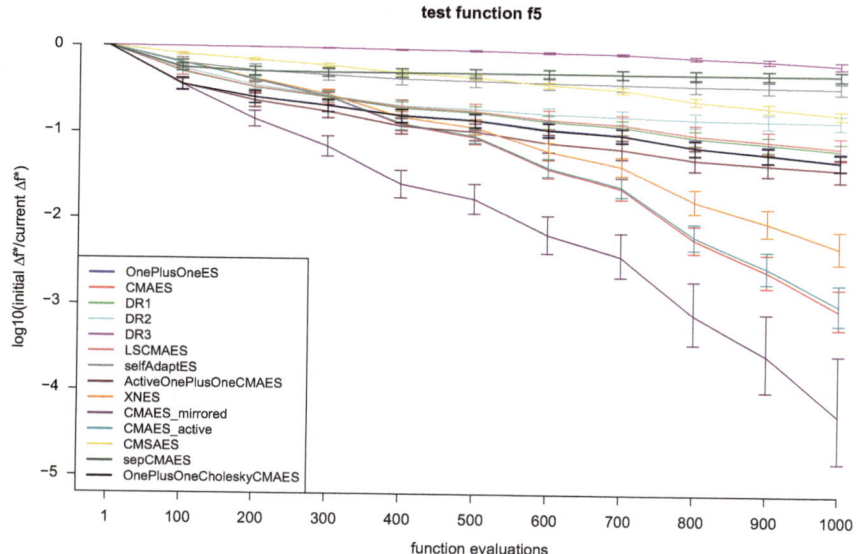

Fig. 4.12 Convergence plot for test function f_5 showing the order of magnitude of fitness value normalized w.r.t. the fitness of the inital search point for the number of function evaluations $\{100, 200, \ldots, 1,000\}$. Error bars reflect the 20% respectively 80% quantiles of the 1,000 conducted runs and the *solid line* represents their mean

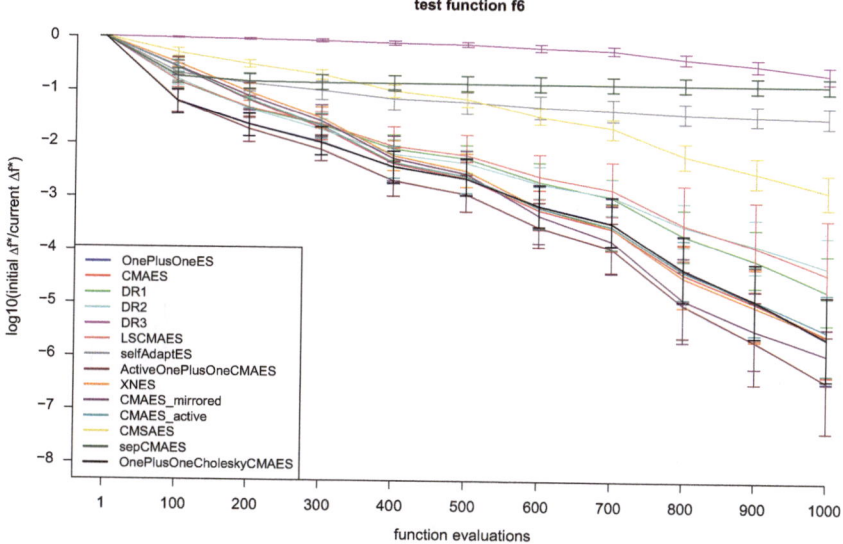

Fig. 4.13 Convergence plot for test function f_6 showing the order of magnitude of fitness value normalized w.r.t. the fitness of the inital search point for the number of function evaluations $\{100, 200, \ldots, 1,000\}$. Error bars reflect the 20% respectively 80% quantiles of the 1,000 conducted runs and the *solid line* represents their mean

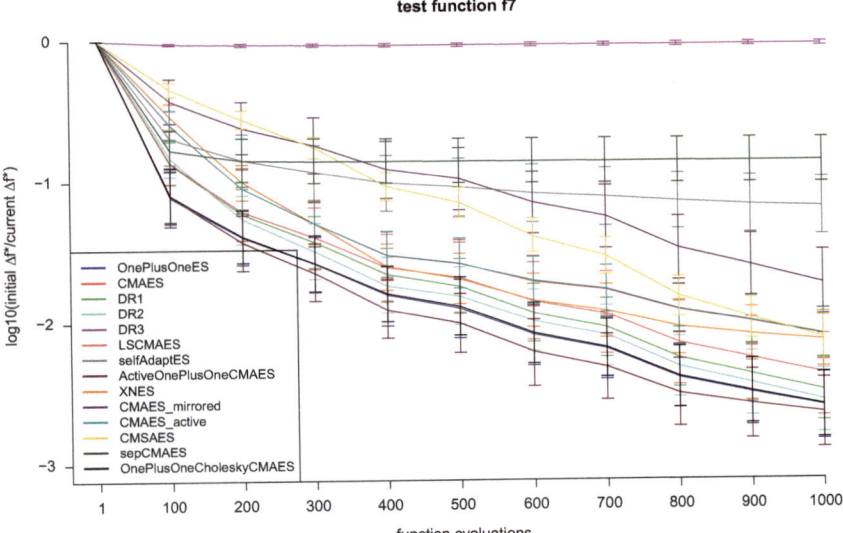

Fig. 4.14 Convergence plot for test function f_7 showing the order of magnitude of fitness value normalized w.r.t. the fitness of the inital search point for the number of function evaluations $\{100, 200, \ldots, 1{,}000\}$. Error bars reflect the 20 % respectively 80 % quantiles of the 1,000 conducted runs and the *solid line* represents their mean

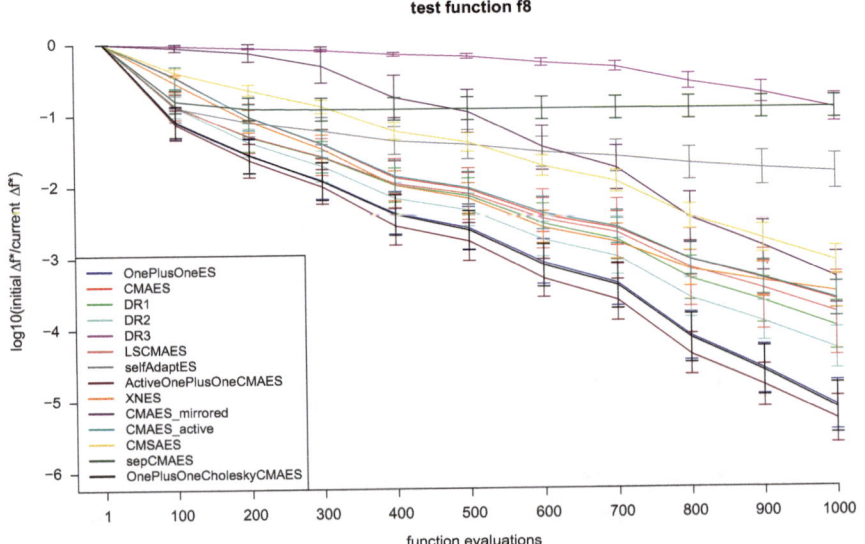

Fig. 4.15 Convergence plot for test function f_8 showing the order of magnitude of fitness value normalized w.r.t. the fitness of the inital search point for the number of function evaluations $\{100, 200, \ldots, 1{,}000\}$. Error bars reflect the 20 % respectively 80 % quantiles of the 1,000 conducted runs and the *solid line* represents their mean

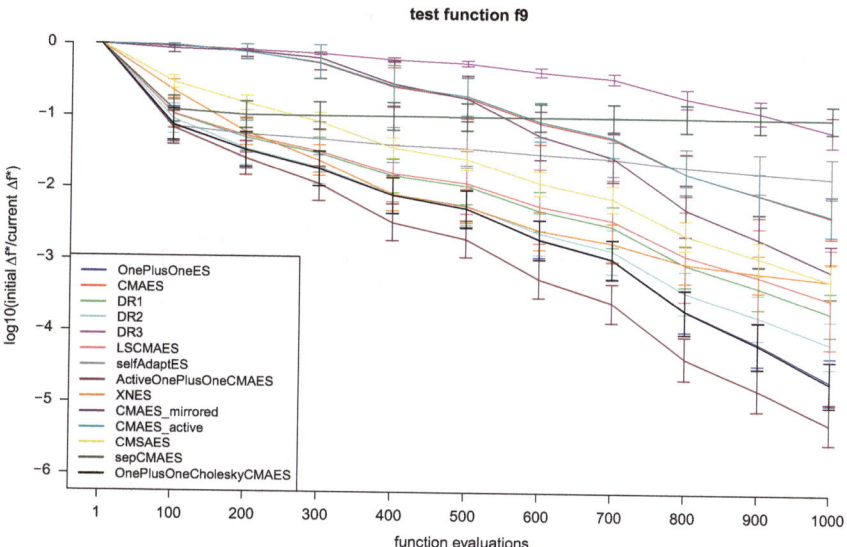

Fig. 4.16 Convergence plot for test function f_9 showing the order of magnitude of fitness value normalized w.r.t. the fitness of the inital search point for the number of function evaluations $\{100, 200, \ldots, 1{,}000\}$. Error bars reflect the 20% respectively 80% quantiles of the $1{,}000$ conducted runs and the *solid line* represents their mean

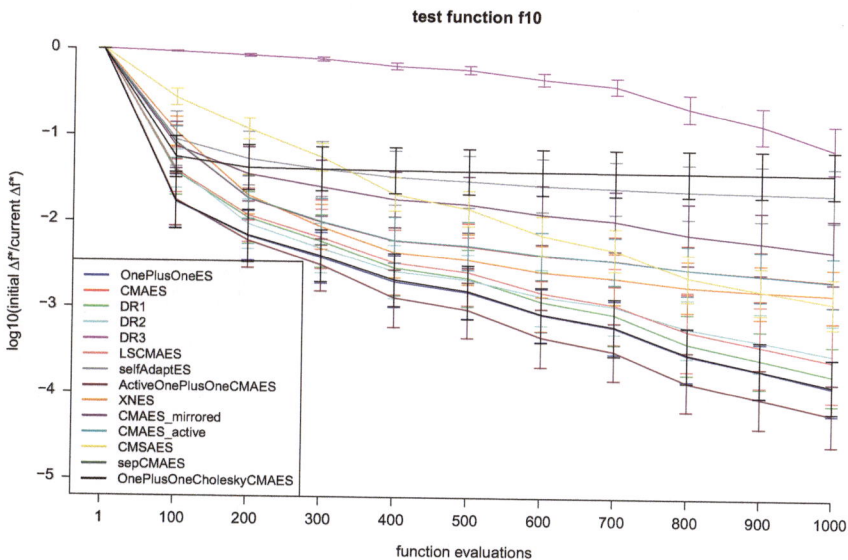

Fig. 4.17 Convergence plot for test function f_{10} showing the order of magnitude of fitness value normalized w.r.t. the fitness of the inital search point for the number of function evaluations $\{100, 200, \ldots, 1{,}000\}$. Error bars reflect the 20% respectively 80% quantiles of the $1{,}000$ conducted runs and the *solid line* represents their mean

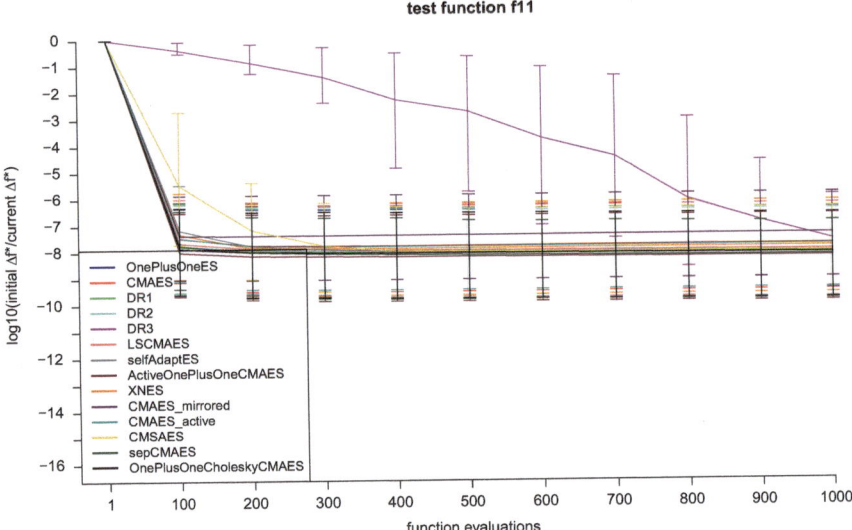

Fig. 4.18 Convergence plot for test function f_{11} showing the order of magnitude of fitness value normalized w.r.t. the fitness of the inital search point for the number of function evaluations $\{100, 200, \ldots, 1,000\}$. Error bars reflect the 20 % respectively 80 % quantiles of the 1,000 conducted runs and the *solid line* represents their mean

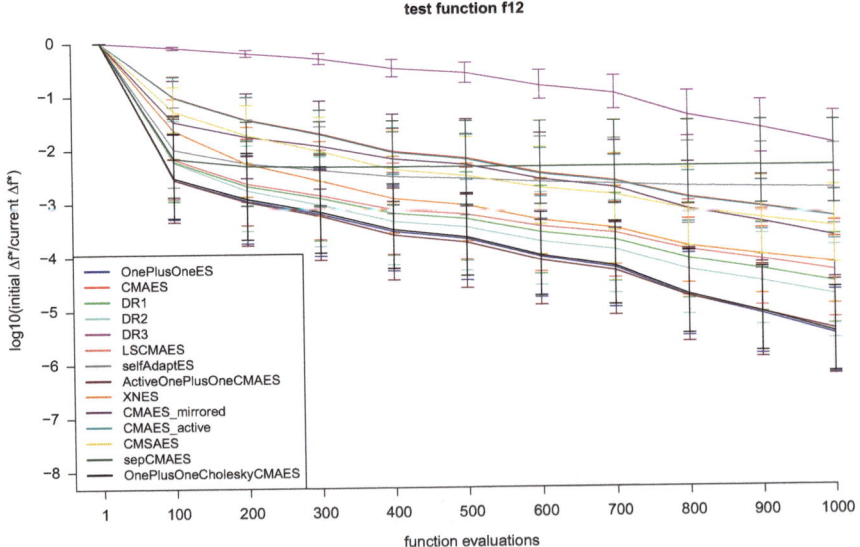

Fig. 4.19 Convergence plot for test function f_{12} showing the order of magnitude of fitness value normalized w.r.t. the fitness of the inital search point for the number of function evaluations $\{100, 200, \ldots, 1,000\}$. Error bars reflect the 20 % respectively 80 % quantiles of the 1,000 conducted runs and the *solid line* represents their mean

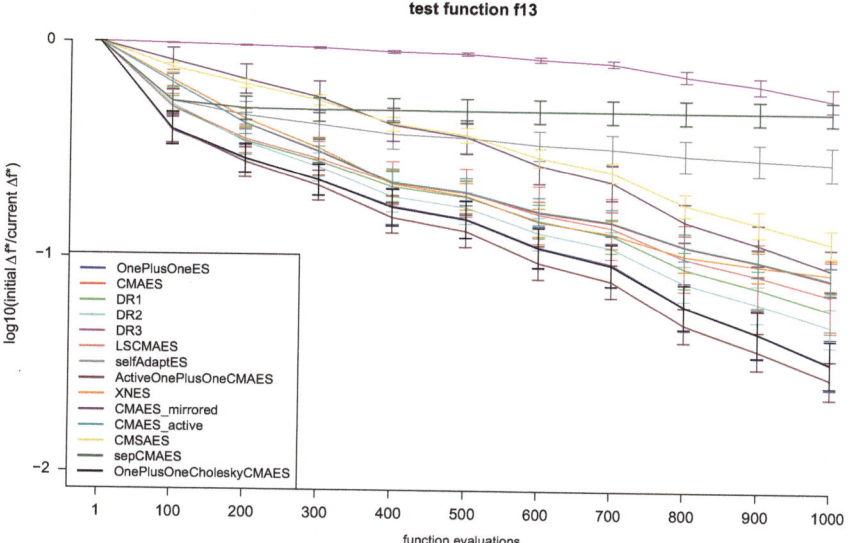

Fig. 4.20 Convergence plot for test function f_{13} showing the order of magnitude of fitness value normalized w.r.t. the fitness of the inital search point for the number of function evaluations $\{100, 200, \ldots, 1{,}000\}$. Error bars reflect the $20\,\%$ respectively $80\,\%$ quantiles of the $1{,}000$ conducted runs and the *solid line* represents their mean

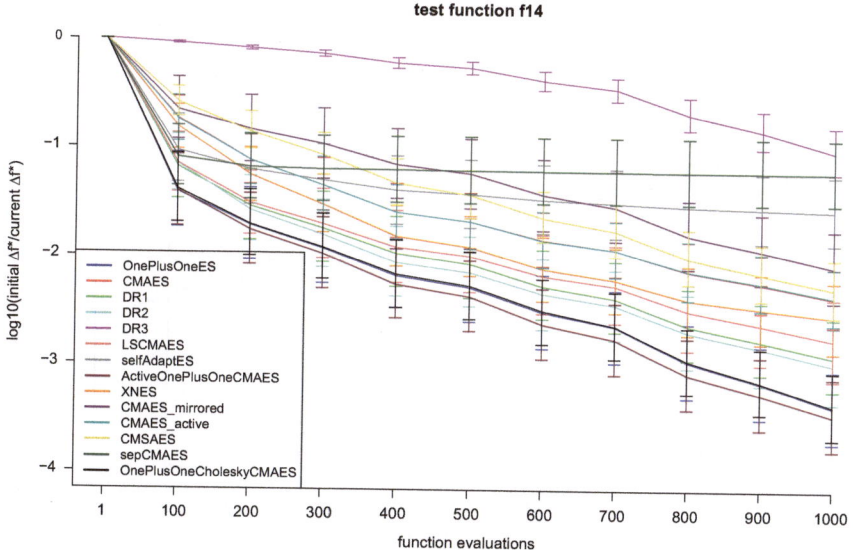

Fig. 4.21 Convergence plot for test function f_{14} showing the order of magnitude of fitness value normalized w.r.t. the fitness of the inital search point for the number of function evaluations $\{100, 200, \ldots, 1{,}000\}$. Error bars reflect the $20\,\%$ respectively $80\,\%$ quantiles of the $1{,}000$ conducted runs and the *solid line* represents their mean

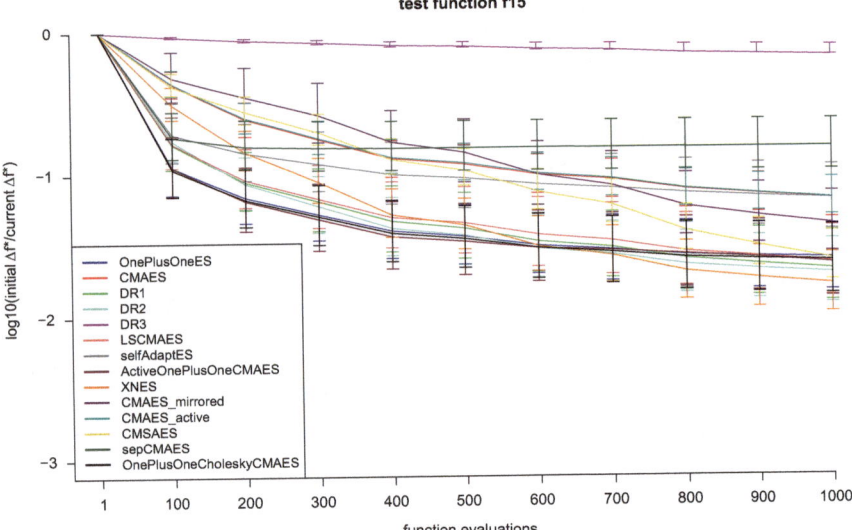

Fig. 4.22 Convergence plot for test function f_{15} showing the order of magnitude of fitness value normalized w.r.t. the fitness of the inital search point for the number of function evaluations $\{100, 200, \ldots, 1{,}000\}$. Error bars reflect the 20 % respectively 80 % quantiles of the 1,000 conducted runs and the *solid line* represents their mean

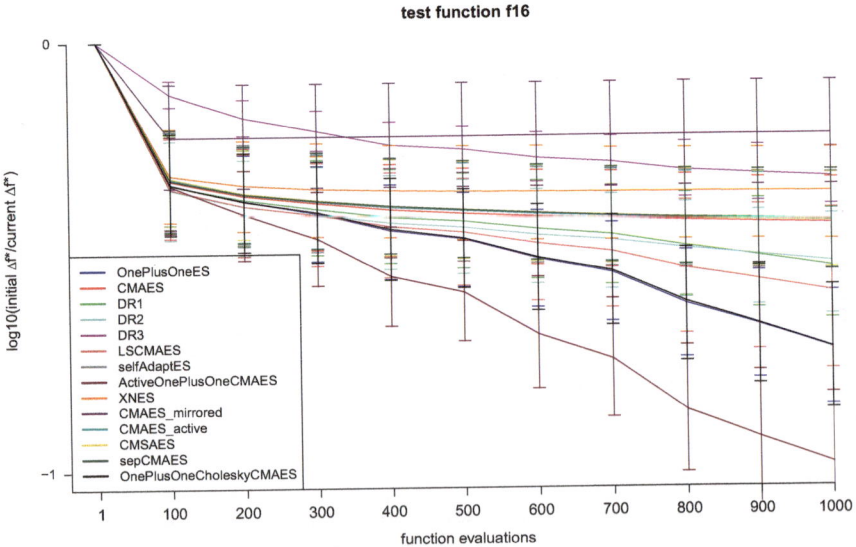

Fig. 4.23 Convergence plot for test function f_{16} showing the order of magnitude of fitness value normalized w.r.t. the fitness of the inital search point for the number of function evaluations $\{100, 200, \ldots, 1{,}000\}$. Error bars reflect the 20 % respectively 80 % quantiles of the 1,000 conducted runs and the *solid line* represents their mean

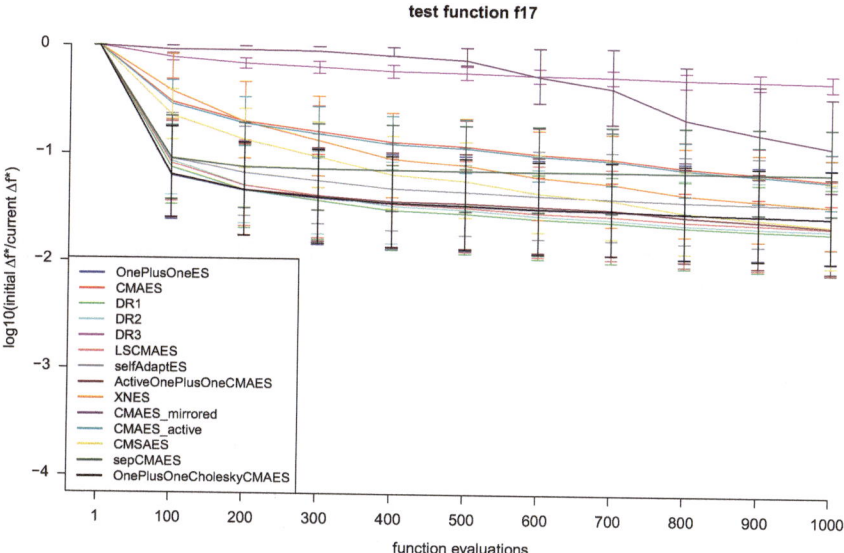

Fig. 4.24 Convergence plot for test function f_{17} showing the order of magnitude of fitness value normalized w.r.t. the fitness of the inital search point for the number of function evaluations $\{100, 200, \ldots, 1{,}000\}$. Error bars reflect the 20% respectively 80% quantiles of the $1{,}000$ conducted runs and the *solid line* represents their mean

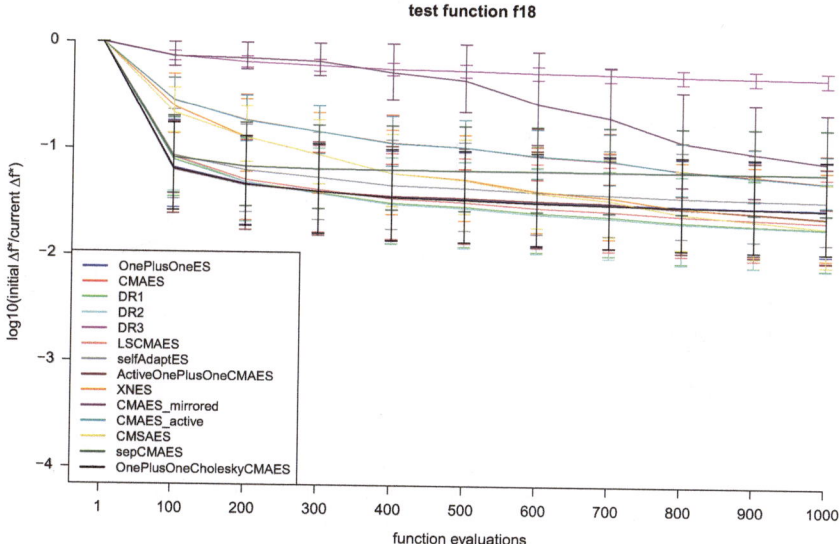

Fig. 4.25 Convergence plot for test function f_{18} showing the order of magnitude of fitness value normalized w.r.t. the fitness of the inital search point for the number of function evaluations $\{100, 200, \ldots, 1{,}000\}$. Error bars reflect the 20% respectively 80% quantiles of the $1{,}000$ conducted runs and the *solid line* represents their mean

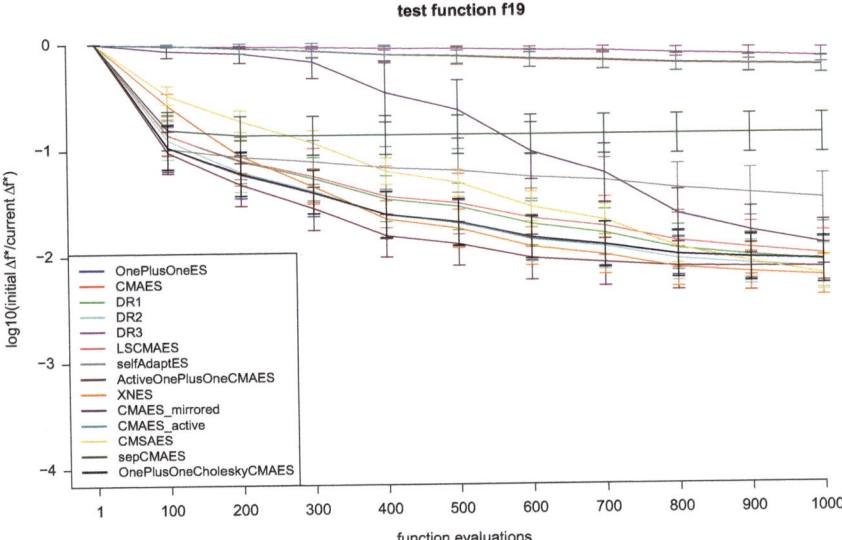

Fig. 4.26 Convergence plot for test function f_{19} showing the order of magnitude of fitness value normalized w.r.t. the fitness of the inital search point for the number of function evaluations $\{100, 200, \ldots, 1{,}000\}$. Error bars reflect the 20 % respectively 80 % quantiles of the 1,000 conducted runs and the *solid line* represents their mean

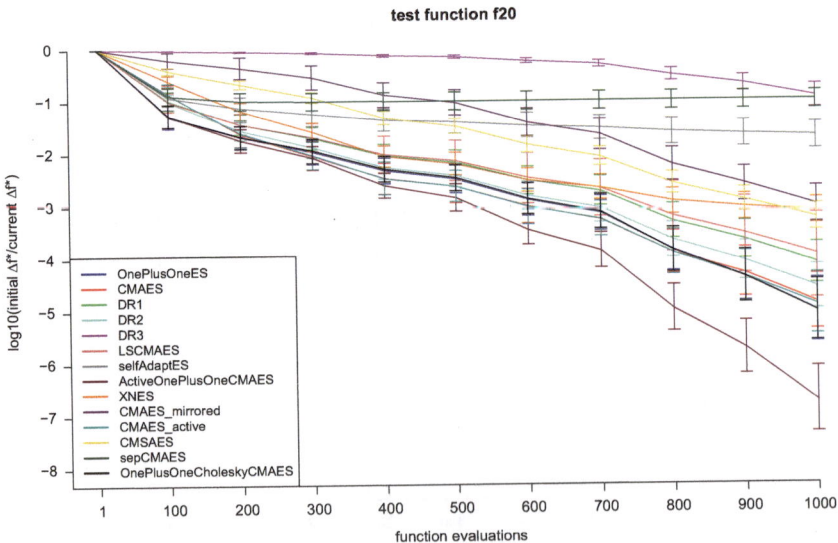

Fig. 4.27 Convergence plot for test function f_{20} showing the order of magnitude of fitness value normalized w.r.t. the fitness of the inital search point for the number of function evaluations $\{100, 200, \ldots, 1{,}000\}$. Error bars reflect the 20 % respectively 80 % quantiles of the 1,000 conducted runs and the *solid line* represents their mean

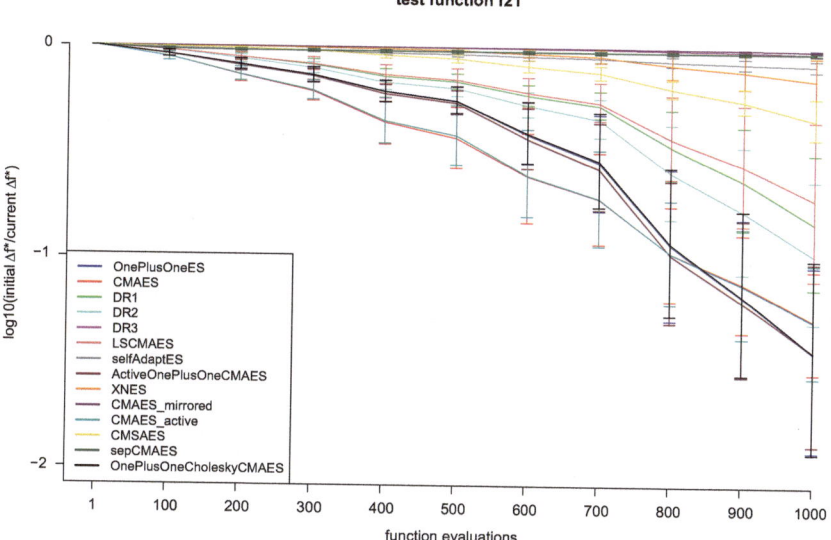

Fig. 4.28 Convergence plot for test function f_{21} showing the order of magnitude of fitness value normalized w.r.t. the fitness of the inital search point for the number of function evaluations $\{100, 200, \ldots, 1{,}000\}$. Error bars reflect the 20% respectively 80% quantiles of the $1{,}000$ conducted runs and the *solid line* represents their mean

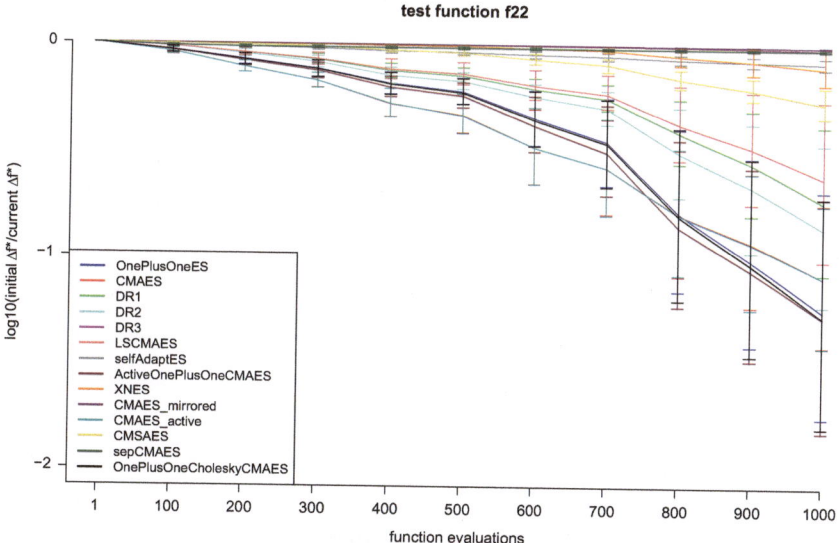

Fig. 4.29 Convergence plot for test function f_{22} showing the order of magnitude of fitness value normalized w.r.t. the fitness of the inital search point for the number of function evaluations $\{100, 200, \ldots, 1{,}000\}$. Error bars reflect the 20% respectively 80% quantiles of the $1{,}000$ conducted runs and the *solid line* represents their mean

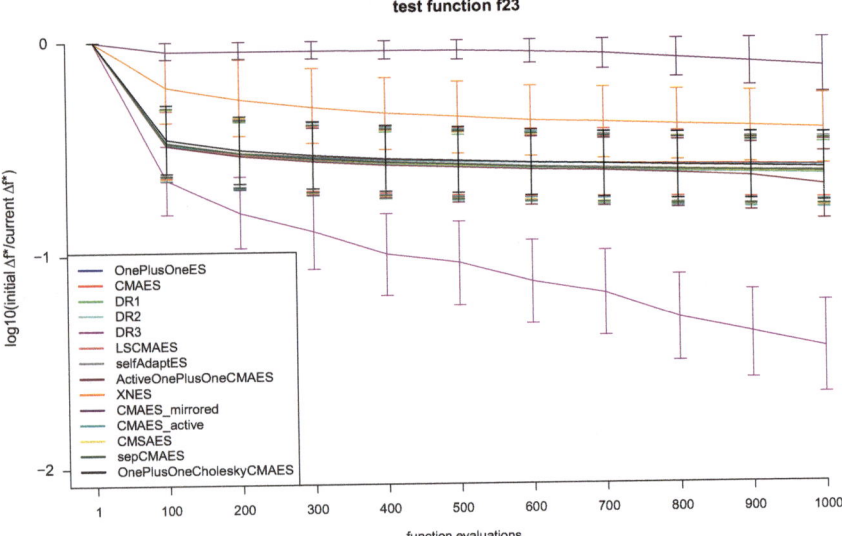

Fig. 4.30 Convergence plot for test function f_{23} showing the order of magnitude of fitness value normalized w.r.t. the fitness of the inital search point for the number of function evaluations $\{100, 200, \ldots, 1,000\}$. Error bars reflect the 20% respectively 80% quantiles of the 1,000 conducted runs and the *solid line* represents their mean

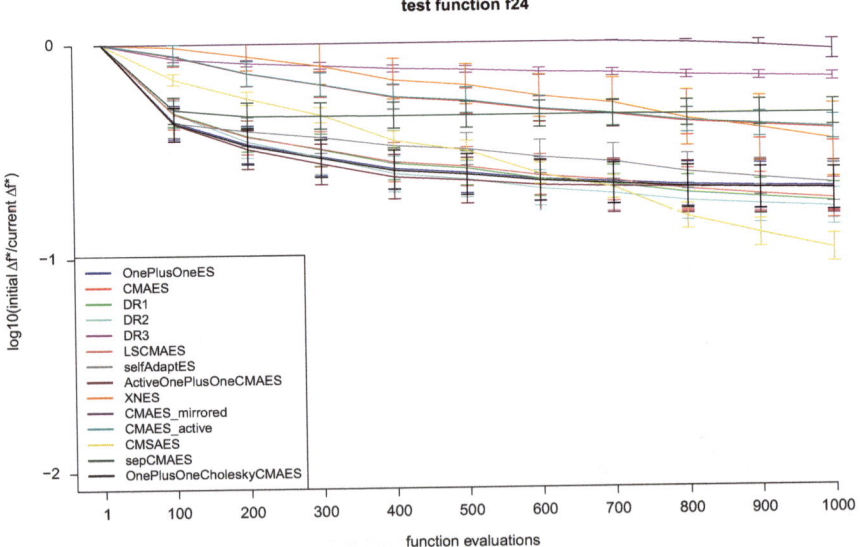

Fig. 4.31 Convergence plot for test function f_{24} showing the order of magnitude of fitness value normalized w.r.t. the fitness of the inital search point for the number of function evaluations $\{100, 200, \ldots, 1,000\}$. Error bars reflect the 20% respectively 80% quantiles of the 1,000 conducted runs and the *solid line* represents their mean

only two cases, for $C_t = 100$. The very simple $(1 + 1)$-ES performs surprisingly well, especially on unimodal functions. On multimodal test functions, the simple DR2 strategy also performs reasonably well, but not for unimodal test functions. Overall, the $(1 + 1)$-Active-CMA-ES is clearly recommendable due to its consistent performance across all functions tested.

Chapter 5
Summary

Contemporary Evolution Strategies (ES) are the subject of this book. Motivated by optimization scenarios with very few function evaluations, experiments to empirically analyze the efficiency of contemporary ES were conducted. Usually such experiments are performed on standardized test functions with many function evaluations using an analysis approach which is not suited to the optimization scenarios considered in this book. We chose another approach that allows ranking of ES algorithms by significant qualitative differences. Our experiments show that there are significant differences regarding the efficiency of ES algorithms even with very few function evaluations. In general the Active-CMA-ES is the best of the algorithms tested with a function evaluation budget of $100n$ for search space dimensionality n. For even smaller budgets like $50n$ and $25n$ several variants of $(1+1)$-strategies turn out to be best, especially on unimodal test functions. To get even closer to a realistic scenario of very small function evaluation budgets, an empirical investigation was conducted for fixed dimensionality of $n = 100$ and a budget of as few as 100, 200, up to 1,000 function evaluations. Based on many real-world optimization projects with objective functions represented by time-consuming simulation programs, often requiring 24 h of computing effort for a single simulation run, this setting represents realistic applications such as occur, e.g., in multidisciplinary design optimization in the automotive industry (see e.g. [22]). As shown in Sect. 4.4, for such a scenario the results are different and the previously winning algorithms, the Active-CMA-ES followed by the (μ_W, λ)-CMA-ES, now have quite mediocre performance. Instead, certain $(1+1)$-strategies clearly outperform population-based approaches, and in particular the $(1+1)$-Active-CMA-ES ranks best in many cases, followed by the $(1+1)$-Cholesky-CMA-ES; see Table 5.1. Interestingly, the DR2 algorithm also performs quite reasonably for the multimodal objective functions; even sometimes outperforming the $(1+1)$-Active-CMA-ES. This result is noticeable from a practical perspective due to the fact that DR2 uses an offspring population size of ten individuals, such that a parallel execution of simulation runs is possible. Moreover, from this point of view the DR2 strategy is the overall best option provided that parallel execution of function evaluations is desired.

T. Bäck et al., *Contemporary Evolution Strategies*, Natural Computing Series,
DOI 10.1007/978-3-642-40137-4_5, © Springer-Verlag Berlin Heidelberg 2013

Table 5.1 Overview of the two best-ranked algorithms by function class for $n = 100$ and very small function evaluation budgets (i.e., up to 1,000 evaluations)

	Unimodal	Multimodal
Separable	1: (1+1)-Active-CMA-ES	1: (1+1)-Active-CMA-ES (for $C_t \leq 700$); DR2 (for $C_t > 700$)
	2: (1+1)-Cholesky-CMA-ES and (1+1)-ES	2: (1+1)-Cholesky-CMA-ES (for $C_t < 700$)
Non-separable	1: (1+1)-Active-CMA-ES	1: (1+1)-Active-CMA-ES
	2: (1+1)-Cholesky-CMA-ES and (1+1)-ES	2: DR2 (for $C_t > 300$)

As clarified by the discussion above, there is no single straightforward answer to the search for the best evolution strategy for challenging practical applications with small function evaluation budget. However, some interesting conclusions can be drawn, and they also indicate quite interesting additional areas for future research, such as the following:

- Constraint handling: Except for simple box constraints defining the feasible region for the parameters of the objective function, no constraints were considered. It is quite important, however, in practical applications, to be able to handle linear and nonlinear constraints on parameters as well as some of the outputs of simulation runs—which would correspond to "black box" constraints deciding on the feasibility of a solution.
- Multiple-criteria optimization: The BBOB test function set deals only with single-objective problems, but many real-world problems are multiple-criteria optimization tasks. It will be very interesting to evaluate the quality of Pareto-front approximations achieved by appropriate algorithmic extensions for multiple criteria optimization, also under limited budgets for the number of objective function evaluations.

The goal of such further investigations would be, again, to enhance our understanding of the real-world problem-solving capabilities of contemporary evolution strategies.

To give an overview of contemporary ES, they are described by their key ideas and by providing the pseudocode of algorithms in Chap. 2. In Chap. 3 the contemporary ES are taxonomically assigned to classes. The main classes are identified according to restart heuristics, methods of covariance adaptation and techniques for avoiding function evaluations.

Bibliography

1. S. Amari, Natural gradient works efficiently in learning. Neural Comput. **10**(2), 251–276 (1998)
2. D.V. Arnold, N. Hansen, Active covariance matrix adaptation for the (1+1)-CMA-ES, in *Proceedings of the 12th Annual Conference on Genetic and Evolutionary Computation (GECCO'10)*, Portland, ed. by M. Pelikan, J. Branke (ACM, New York, 2010), pp. 385–392
3. D.V. Arnold, R. Salomon, Evolutionary gradient search revisited. IEEE Trans. Evol. Comput. **11**(4), 480–495 (2007)
4. A. Auger, N. Hansen, Performance evaluation of an advanced local search evolutionary algorithm, in *Proceedings of the IEEE Congress on Evolutionary Computation (CEC'05)*, Edinburgh, vol. 2, ed. by B. McKay et al. (IEEE, Piscataway, 2005), pp. 1777–1784
5. A. Auger, N. Hansen, A restart CMA evolution strategy with increasing population size, in *Proceedings of the IEEE Congress on Evolutionary Computation (CEC'05)*, Edinburgh, vol. 2, ed. by B. McKay et al. (IEEE, Piscataway, 2005), pp. 1769–1776
6. A. Auger, M. Schoenauer, N. Vanhaecke, LS-CMA-ES: a second-order algorithm for covariance matrix adaptation, in *Proceedings of the 8th International Conference on Parallel Problem Solving from Nature (PPSN VIII)*, Birmingham, ed. by X. Yao et al. Volume 3242 of Lecture Notes in Computer Science (Springer, Berlin, 2004), pp. 182–191
7. A. Auger, D. Brockhoff, N. Hansen, Mirrored sampling in evolution strategies with weighted recombination, in *Proceedings of the 13th Annual Genetic and Evolutionary Computation Conference (GECCO'11)*, Dublin, ed. by N. Krasnogor, P.L. Lanzi (ACM, New York, 2011), pp. 861–868
8. T. Bäck, *Evolutionary Algorithms in Theory and Practice* (Oxford University Press, New York, 1996)
9. T. Bäck, D.B. Fogel, Z. Michalewicz, *Evolutionary Computation 1: Basic Algorithms and Operators* (Taylor & Francis, New York, 2000)
10. T. Bäck, D.B. Fogel, Z. Michalewicz, *Evolutionary Computation 2: Advanced Algorithms and Operators*. Evolutionary Computation (Taylor & Francis, New York, 2000)
11. T. Bartz-Beielstein, C. Lasarczyk, M. Preuss, Sequential parameter optimization, in *Proceedings of the IEEE Congress on Evolutionary Computation (CEC'05)*, Edinburgh, ed. by B. McKay et al. (IEEE, Piscataway, 2005), pp. 773–780
12. N. Beume, B. Naujoks, M. Emmerich, SMS-EMOA: multiobjective selection based on dominated hypervolume. Eur. J. Oper. Res. **181**, 1653–1669 (2007)
13. H.-G. Beyer, B. Sendhoff, Covariance matrix adaptation revisited – the CMSA evolution strategy, in *Proceedings of the 10th International Conference on Parallel Problem Solving from Nature (PPSN X)*, Dortmund, ed. by G. Rudolph et al. Volume 5199 in Lecture Notes in Computer Science (Springer, Berlin, 2008), pp. 123–132

T. Bäck et al., *Contemporary Evolution Strategies*, Natural Computing Series, DOI 10.1007/978-3-642-40137-4, © Springer-Verlag Berlin Heidelberg 2013

14. Z. Bouzarkouna, A. Auger, D.-Y. Ding, Investigating the local-meta-model CMA-ES for large population sizes, in *Proceedings of the 3rd European Event on Bioinspired Algorithms for Continuous Parameter Optimisation (EvoNUM'10)*, Istanbul, Turkey, ed. by C. Di Chio et al. Volume 6024 in Lecture Notes in Computer Science (Springer, Berlin, 2010), pp. 402–411

15. Z. Bouzarkouna, A. Auger, D.-Y. Ding, Local-meta-model CMA-ES for partially separable functions, in *Proceedings of the 13th Annual Genetic and Evolutionary Computation Conference (GECCO'11)*, Dublin, ed. by N. Krasnogor et al. (ACM, New York, 2011), pp. 869–876

16. D. Brockhoff, A. Auger, N. Hansen, D.V. Arnold, T. Hohm, Mirrored sampling and sequential selection for evolution strategies, in *Proceedings of the 11th International Conference on Parallel Problem Solving from Nature (PPSN XI)*, Kraków, ed. by R. Schaefer et al. Volume 6238 in Lecture Notes in Computer Science. (Springer, Berlin, 2010), pp. 11–21

17. I.N. Bronstein, K.A. Semendjajew, G. Musiol, H. Muehlig, *Taschenbuch der Mathematik*, 7th edn. (Harri Deutsch, Frankfurt am Main, 2008)

18. C.A. Coello Coello, Constraint-handling techniques used with evolutionary algorithms, in *Proceedings of the 13th Annual Genetic and Evolutionary Computation Conference (GECCO'11), Companion Material*, Dublin, ed. by N. Krasnogor et al. (ACM, New York, 2011), pp. 1137–1160

19. C. Darwin, *On the Origin of Species by Means of Natural Selection: Or, The Preservation of Favoured Races in the Struggle for Life* (J. Murray, London, 1860)

20. K. Deb, *Multiobjective Optimization Using Evolutionary Algorithms*. Wiley-Interscience Series in Systems and Optimization (Wiley, Chichester, 2001)

21. K. Deb, A. Pratap, S. Agarwal, T. Meyarivan, A fast and elitist multiobjective genetic algorithm: NSGA-II. IEEE Trans. Evol. Comput. **6**(2), 182–197 (2002)

22. F. Duddeck, Multidisciplinary optimization of car bodies. Struct. Multidiscip. Optim. **35**(4), 375–389 (2008)

23. J.W. Eaton, *GNU Octave Manual* (Network Theory Limited, Godalming, 2002)

24. A.E. Eiben, M. Jelasity, A critical note on experimental research methodology in EC, in *Proceedings of the 2002 Congress on Evolutionary Computation (CEC'02)*, Honolulu, ed. by R. Eberhart et al. (IEEE, Piscataway, 2002), pp. 582–587

25. A.E. Eiben, J.E. Smith, *Introduction to Evolutionary Computing*. Natural Computing Series (Springer, Berlin, 2003)

26. T. Glasmachers, T. Schaul, Y. Sun, D. Wierstra, J. Schmidhuber, Exponential natural evolution strategies, in *Proceedings of the 12th Annual Conference on Genetic and Evolutionary Computation (GECCO'10)*, Portland, ed. by M. Pelikan, J. Branke (ACM, New York, 2010)

27. D.E. Goldberg, *Genetic Algorithms in Search, Optimization, and Machine Learning* (Addison-Wesley, Boston, 1989)

28. W.H. Greene, *Econometric Analysis*, 4th edn. (Prentice Hall, Upper Saddle River, 1997)

29. N. Hansen, The CMA evolution strategy: a tutorial. Continuously updated technical report, available via http://www.lri.fr/~hansen/cmatutorial.pdf. Accessed 12 Mar 2011

30. N. Hansen, S. Kern, Evaluating the CMA evolution strategy on multimodal test functions, in *Proceedings of the 9th International Conference on Parallel Problem Solving from Nature (PPSN VIII)*, Birmingham. Volume 3242 of Lecture Notes in Computer Science, ed. by X. Yao et al. (Springer, 2004), pp. 282–291

31. N. Hansen, A. Ostermeier, Adapting arbitrary normal mutation distributions in evolution strategies: the covariance matrix adaptation, in *Proceedings of the 1996 IEEE International Conference on Evolutionary Computation (ICEC'96)*, Nagoya, ed. by Y. Davidor et al. (IEEE, Piscataway, 1996), pp. 312–317

32. N. Hansen, A. Ostermeier, Completely derandomized self-adaptation in evolution strategies. Evol. Comput. **9**(2), 159–195 (2001)

33. N. Hansen, A. Ostermeier, A. Gawelczyk, On the adaptation of arbitrary normal mutation distributions in evolution strategies: the generating set adaptation, in *Proceedings of the 6th International Conference on Genetic Algorithms (ICGA 6)*, Pittsburgh, ed. by L.J. Eshelman (Morgan Kaufmann, San Francisco, 1995), pp. 57–64

34. N. Hansen, A. Auger, S. Finck, R. Ros, Real-parameter black-box optimization benchmarking 2010: experimental setup. Research report RR-7215, INRIA, 2010
35. N. Hansen, A. Auger, R. Ros, S. Finck, P. Posik, Comparing results of 31 algorithms from the black-box optimization benchmarking BBOB-2009, in *Proceedings of the 12th International Conference on Genetic and Evolutionary Computation Conference (GECCO'10), Companion Material*, Portland, ed. by M. Pelikan, J. Branke (ACM, New York, 2010), pp. 1689–1696
36. J. Hartung, B. Elpelt, K.H. Klösener, *Statistik*, 14th edn. (Oldenbourg, München, 2005)
37. T. Hastie, R. Tibshirani, J.H. Friedman, *The Elements of Statistical Learning: Data Mining, Inference, and Prediction*, 2nd edn. Springer Series in Statistics (Springer, Berlin, 2009)
38. C. Igel, T. Suttorp, N. Hansen, A computational efficient covariance matrix update and a $(1+1)$-CMA for evolution strategies, in *Proceedings of the 8th Annual Conference on Genetic and Evolutionary Computation (GECCO'06)*, Seattle, ed. by M. Keijzer et al. (ACM, New York, 2006), pp. 453–460
39. G.A. Jastrebski, Improving evolution strategies through active covariance matrix adaptation. Master's thesis, Faculty of Computer Science, Dalhousie University, 2005
40. G.A. Jastrebski, D.V. Arnold, Improving evolution strategies through active covariance matrix adaptation, in *Proceedings of the 2006 IEEE Congress on Evolutionary Computation (CEC'06)*, Vancouver, BC, Canada, ed. by G.G. Yen et al. (IEEE, Piscataway, 2006), pp. 2814–2821
41. S. Kern, N. Hansen, P. Koumoutsakos, Local meta-models for optimization using evolution strategies, in *Proceedings of the 9th International Conference on Parallel Problem Solving from Nature (PPSN IX)*, Reykjavik, ed. by T.P. Runarsson et al. (Springer, Berlin, 2006), pp. 939–948
42. O. Kramer, A review of constraint-handling techniques for evolution strategies. Appl Comput. Int. Soft Comput. **2010**, 1–11 (2010)
43. R. Li, Mixed-integer evolution strategies for parameter optimization and their applications to medical image analysis. PhD thesis, Leiden Institute of Advanced Computer Science (LIACS), Faculty of Science, Leiden University, 2009
44. D.G. Luenberger, Y. Ye, *Linear and Nonlinear Programming*, 2nd edn. (Springer, Berlin, 2003)
45. T.M. Mitchell, *Machine Learning* (McGraw-Hill, New York, 1997)
46. S.D. Müller, N. Hansen, P. Koumoutsakos, Increasing the serial and the parallel performance of the CMA-evolution strategy with large populations, in *Proceedings of the 7th International Conference on Parallel Problem Solving from Nature (PPSN VII)*, Granada, ed. by J.J. Merelo et al. Volume 2439 of Lecture Notes in Computer Science (Springer, Berlin, 2002), pp. 422–431
47. A. Ostermeier, A. Gawelczyk, N. Hansen, A derandomized approach to self adaptation of evolution strategies. Evol. Comput. **2**(4), 369–380 (1994)
48. A. Ostermeier, A. Gawelczyk, N. Hansen, Step-size adaptation based on non-local use of selection information, in *Proceedings of the 3rd International Conference on Parallel Problem Solving from Nature (PPSN III)*, Jerusalem, ed. by Y. Davidor et al. Volume 866 of Lecture Notes in Computer Science (Springer, Berlin, 1994), pp. 189–198
49. K.V. Price, Differential evolution vs. the functions of the second ICEO, in *Proceedings of the IEEE International Congress on Evolutionary Computation*, Indianapolis, ed. by B. Porto et al. (IEEE, Piscataway, 1997), pp. 153–157
50. R Development Core Team, *R: A Language and Environment for Statistical Computing*. R Foundation for Statistical Computing, Vienna, 2011. ISBN:3-900051-07-0
51. I. Rechenberg, Cybernetic solution path of an experimental problem. Royal Aircraft Establishment, Library Translation 1122, Farnborough, 1965
52. I. Rechenberg, *Evolutionsstrategie: Optimierung Technischer Systeme nach Prinzipien der biologischen Evolution* (Frommann-Holzboog, Stuttgart, 1973)
53. I. Rechenberg, *Evolutionsstrategie'94* (Frommann-Holzboog, Stuttgart, 1994)
54. R. Ros, N. Hansen, A simple modification in CMA-ES achieving linear time and space complexity, in *Proceedings of the 10th International Conference on Parallel Problem Solving from Nature (PPSN X)*, Dortmund, ed. by G. Rudolph et al. Volume 5199 of Lecture Notes in Computer Science (Springer, Berlin, 2008), pp. 296–305

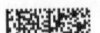

55. G. Rudolph, On correlated mutations in evolution strategies, in *Proceedings of the 2nd International Conference on Parallel Problem Solving from Nature (PPSN II)*, Brussels, ed. by R. Männer, B. Manderick (Elsevier, Amsterdam, 1992), pp. 105–114

56. G. Rudolph, An evolutionary algorithm for integer programming, in *Proceedings of the 3rd Conference on Parallel Problem Solving from Nature (PPSN III)*, Jerusalem, ed. by Y. Davidor et al. Volume 866 of Lecture Notes in Computer Science (Springer, Berlin, 1994), pp. 63–66

57. G. Rudolph, *Convergence Properties of Evolutionary Algorithms* (Kovač, Hamburg, 1997)

58. G. Rudolph, Evolutionary strategies, in *Handbook of Natural Computing*, ed. by G. Rozenberg, T. Bäck, J.N. Kok (Springer, Berlin, 2012)

59. H.-P. Schwefel, Kybernetische Evolution als Strategie der experimentellen Forschung in der Strömungstechnik. Diplomarbeit, Technische Universität Berlin, Hermann Föttinger–Institut für Strömungstechnik, 1964

60. H.-P. Schwefel, Evolutionsstrategie und numerische Optimierung. Dr.-Ing. Dissertation, Fachbereich Verfahrenstechnik, Technische Universität Berlin, 1975

61. H.-P. Schwefel, *Numerische Optimierung von Computer-Modellen Mittels der Evolutionsstrategie* (Birkhäuser, Basel, 1977)

62. H.-P. Schwefel, *Numerical Optimization of Computer Models* (Wiley, Chichester, 1981)

63. O.M. Shir, *Niching in Derandomized Evolution Strategies and its Applications in Quantum Control*. PhD thesis, University of Leiden, The Netherlands, 2008

64. A. Stuart, K. Ord, S. Arnold, *Kendall's Advanced Theory of Statistics, Classical Inference and the Linear Model*. Volume 2 in Kendall's Library of Statistics (Wiley, Chichester, 2009)

65. P.N. Suganthan, N. Hansen, J.J. Liang, K. Deb, Y.P. Chen, A. Auger, S. Tiwari, Problem definitions and evaluation criteria for the CEC 2005 special session on real-parameter optimization. Technical Report 2005005, Nanyang Technological University, Singapore and KanGAL (Kanpur Genetic Algorithms Laboratory, IIT Kanpur), 2005

66. Y. Sun, D. Wierstra, T. Schaul, J. Schmidhuber, Efficient natural evolution strategies, in *Proceedings of the 11th Annual Conference on Genetic and Evolutionary Computation (GECCO'09)*, Shanghai, ed. by F. Rothlauf et al. (ACM, New York, 2009), pp. 539–546

67. Y. Sun, D. Wierstra, T. Schaul, J. Schmidhuber, Stochastic search using the natural gradient, in *Proceedings of the 26th Annual International Conference on Machine Learning, ICML'09*, Montreal, ed. by A. Pohoreckyj Danyluk et al. (ACM, New York, 2009), pp. 1161–1168

68. B. Tang, Orthogonal array-based latin hypercubes. J. Am. Stat. Assoc. **88**(424), 1392–1397 (1993)

69. B.L. Welch, The generalization of "Student's" problem when several different population variances are involved. Biometrika **34**, 28–35 (1947)

70. S. Wessing, M. Preuss, G. Rudolph, When parameter tuning actually is parameter control, in *Proceedings of the 13th Annual Conference on Genetic and Evolutionary Computation (GECCO'11)*, Dublin, ed. by N. Krasnogor et al. (ACM, New York, 2011), pp. 821–828

71. D. Wierstra, T. Schaul, J. Peters, J. Schmidhuber, Natural evolution strategies, in *Proceedings of the IEEE Congress on Evolutionary Computation (CEC'08)*, Hong Kong (IEEE, Piscataway, 2008), pp. 3381–3387